2284

This book is part of the Peter Lang Education list.
Every volume is peer reviewed and meets
the highest quality standards for content and production.

PETER LANG
New York • Bern • Frankfurt • Berlin
Brussels • Vienna • Oxford • Warsaw

Seymour W. Itzkoff

2284

World Society, Iaian Vernier's Memoir

PETER LANG
New York • Bern • Frankfurt • Berlin
Brussels • Vienna • Oxford • Warsaw

Library of Congress Cataloging-in-Publication Data
Names: Itzkoff, Seymour W., author.
Title: 2284: world society: Iaian Vernier's memoir / Seymour W. Itzkoff.
Other titles: Twenty-two eighty-four | Two thousand two hundred eighty-four
Description: New York: Peter Lang, 2016.
Identifiers: LCCN 2016001516 | ISBN 978-1-4331-3397-8 (paperback)
ISBN 978-1-4539-1838-8 (ebook)
Subjects: LCSH: Human evolution—Fiction. | Social evolution—Fiction.
Peace—Fiction. | GSAFD: Utopian fiction. | Science fiction.
Classification: LCC PS3609.T88 A617 2016 | DDC 813/.6—dc23
LC record available at http://lccn.loc.gov/2016001516

Bibliographic information published by **Die Deutsche Nationalbibliothek**.
Die Deutsche Nationalbibliothek lists this publication in the "Deutsche
Nationalbibliografie"; detailed bibliographic data are available
on the Internet at http://dnb.d-nb.de/.

© 2016 Peter Lang Publishing, Inc., New York
29 Broadway, 18th floor, New York, NY 10006
www.peterlang.com

All rights reserved.
Reprint or reproduction, even partially, in all forms such as microfilm,
xerography, microfiche, microcard, and offset strictly prohibited.

Copy of marble portrait of Plato by Silanion, c. 370 BCE

Plato, c.427-347 BCE

Plato searched for a concept of justice applicable to the individual as well as to his fellow Hellenes.

Our task is more universal, peace on earth, the democratic life, social equality. Our method must be undergirded by secular, scientific rationality. But, as with Plato, it must be larded with a significant dose of skepticism.

Yet, the utopian quest must continue.

BOOKS BY SEYMOUR W. ITZKOFF
Smith College, Emeritus Professor

Cultural Pluralism and American Education	1969
Ernst Cassirer: Scientific Knowledge and the Concept of Man	1971, 1997 (2nd edition)
A New Public Education	1976
Ernst Cassirer, Philosopher of Culture	1977
Emanuel Feuermann, Virtuoso	1979, 1995 (2nd edition)

The Evolution of Human Intelligence
    1 *The Form of Man, The Evolutionary Origins of Human Intelligence*   1983
    2 *Triumph of the Intelligent, The Creation of Homo sapiens sapiens*   1985
    3 *Why Humans Vary in Intelligence*   1987
    4 *The Making of the Civilized Mind*   1990

How We Learn to Read	1986
Human Intelligence and National Power	1991
The Road to Equality, Evolution and Social Reality	1992
The Decline of Intelligence in America, A Strategy for National Renewal	1994
Children Learning to Read, A Guide for Parents and Teachers	1996

The Human Prospect
    1 *The Inevitable Domination by Man, An Evolutionary
       Detective Story*   2000
    2 *2050: The Collapse of the Global Techno-Economy*   2003
    3 *Intellectual Capital in Twenty-First-Century Politics*   2003
    4 *Rebuilding Western Civilization, Beyond the
       Twenty-First-Century Collapse*   2005

Who Are the Jews?
    1 *Soul of the Israelites*   2004
    2 *A Nation of Philosophers*   2004
    3 *Fatal Gift, Jewish Intelligence and Western Civilization*   2006

The World Energy Crisis and the Task of Retrenchment   2008
The End of Economic Growth: What Does It Mean for American Society?   2009
Judaism's Promise, Meeting the Challenge of Modernity   2014
Liberty's Dilemma: America, Two Nations Dependent/Independent   2014
Humanity's Evolutionary Destiny: A Darwinian Perspective   2016

TABLE OF CONTENTS

Introductory Biographical Note   xiii
Personal Note from Iaian Vernier   xv

Chapter 1.   A Revolutionary Vision   1
Chapter 2.   Into the Tunnel   7
Chapter 3.   Coming Apart   15
Chapter 4.   Light at the End   23
Chapter 5.   Reconstruction   31
Chapter 6.   Brain Power   37
Chapter 7.   Road to Social Equality   43
Chapter 8.   The Long Road Traveled   49
Chapter 9.   *Homo sapiens sapiens*: Evolution's Mystery   57
Chapter 10. Goal of a Unified Humanity   63
Chapter 11. A Work in Progress   69
Chapter 12. Universal and Plural   75
Chapter 13. Ethnicity   81
Chapter 14. Our Democracy   89
Chapter 15. Why Democracy?   97
Chapter 16. The Democratic Life: Criteria   103

Chapter 17. Economic Equality	113
Chapter 18. Economic Progress Report	119
Chapter 19. Perennial Religion	127
Chapter 20. The Fine Arts	137
Chapter 21. Our Esthetic Vision	145
Chapter 22. World Without War	153
Chapter 23. Education	161
Chapter 24. Human Sexuality	167
Chapter 25. Perspective	173
Chapter 26. Final Thought	177

INTRODUCTORY BIOGRAPHICAL NOTE

Iaian Vernier was born in 2204 in what was once called Belgium. His parents were of international heritage and worked in various civil service positions in Europe, Africa and in South America. His studies in Louvain, Wallonia; Cambridge, England; and Heidelberg, Germany were in political theory and philosophy, with a mathematical and legal sub-specialization. Early on, he joined the Secretariat of the World Society shortly after its move to Africa and rose up in its ranks and responsibilities as a very prescient and tough evaluator of plans and process.

Iaian Vernier died in 2294, about one hundred years ago. His final retirement from the World Society Secretariat took place in 2282. His normal retirement at age 70 occurred in 2274. But, because of a serious dispute between medical scientists in Johannesburg, Transvaal Nationality and Nanjing, East Yangtze Nationality, which had enormous potential for increasing human health and life span involving autoimmune therapies, he was called back to service to help adjudicate the dispute for presentation to the Congress. Each of the contesting national entities if confirmed would have been open to special dispensations in population and productivity increases for their achievements. The issue was peacefully resolved, as we all know, and the world is the beneficiary. After four years of additional service on other matters of international concern, Dr. Vernier resumed his well-earned retirement.

Our belief is that we learn from our achievements but even more from our errors of perspective, planning, and applications to the life and destiny of our human cohort. This also applies to the larger tree of life once and today to have existed on this planet. This memoir does not necessarily reflect our current planning perspectives, but does throw light onto our thinking at the time of his writing. Dr. Vernier, in his forty-three years of service, taught us much. To the extent that the changes we have made in the structure of law in these subsequent decades have been successful, to the extent of our past skepticism about our ultimate destiny, we are indebted to Dr. Vernier's probing intellect.

Secretariat, World Society, Nairobi Enclave, 2394

PERSONAL NOTE FROM IAIAN VERNIER

To help the reader understand where I come from, I must set down a few biographical details of my life and work. I am a civil servant, and the child of civil servants. While born and raised mostly on the European continent, my heritage and that of my parents are transnational. Racially and ethnically I represent several continents as does my wife. My heritage is European, Asian and African. My wife comes from an ancient line of North Americans with European origins as well as what has been called Latino.

Our two children have married Asiatic heritage individuals, and our grandchildren have also married individuals with Asiatic and African roots. With our encouragement, I think, we now have great grandchildren on the way. Our family represents our new world citizenry wherein old racial and ethnic origins are rapidly being displaced by new and very different identities. Although our children have adopted our civil service traditions of professional commitment, I believe our grandchildren, having been given wide educational choices, are poised to take on roles in independent artistic and technological institutions.

My one great regret is not having had the perspective of those who live, work, and think outside of governmental domains. I served World Society institutions for forty-three years. Especially gratifying was the opportunity to experience at an early stage in the move of this important institution for peace and prosperity to Nairobi in Africa. For well over a century now the alteration of the human landscape in that continent has been one of our most gratifying experiences, testimony to the wisdom of our predecessors in world government.

After our retirement in 2274, after thirty-nine years of service, my wife and I flounced around the world for four years following the lives and careers of our children and the development of our grandchildren. Then came a new crisis, and I went back to Nairobi trying to help, staying on to give some perspective to my replacements. My devoted spouse, a few years younger than I, wanted to settle down in one place. I especially needed to gain some perspective on what I had inherited in terms of my professional commitments and what our cohort had initiated as a task for long-term world stability.

We decided to look for a new and quiet ecology and nationality for our final years. Having just experienced four years of travel we decided on what has been called northern New England/New Hampshire. We found a relatively small private country residence on which we did some minor renovations. It is surrounded by maples and pines and lies beyond a lake where people for hundreds of years have sailed their boats quietly and in harmony with nature and humanity.

I need to comment on our economic position for it reflects on the power and privilege of those of us who have been responsible for furthering this great revolutionary experiment in worldwide peace and stability. At the least we are economically 'comfortable' in our retirement. Our salaries were always decent; my retirement income is sufficient, but we are not wealthy. In past centuries too many governmental employees have used their latent powers and notoriety to fatten their purse by lobbying and speechifying, garnering enormous emoluments. The services to the constituencies they supposedly represented were essentially covers for obtaining personal wealth and power.

The basis for this new political/economic structure of governance that we are attempting to establish is that public service means just that. An individual should leave public service no more enriched or powerful than when he or she entered such service. Our international laws now require such moral commitments. And, of course we have had to take legal actions against those who have given way to temptation. Such malfeasance is now being paid for behind bars.

As for the weather in this part of 'New England' it has fulfilled the prediction as being bracing; the citizenry are disciplined and not highly charged. But then again, the rate of social change we are attempting to establish worldwide is much slower as compared with earlier eras, the twentieth and twenty-first centuries, for example. It is a bonus that the Secretariat has established for responsibly positioned civil servants that they can remain anonymous. We are able to use a pseudonym to protect our own and our family's privacy. Our way of life, our standard of living, supports this modest quietism that we seek for our 'golden years.'

At age eighty and seventy-six, respectively, and in good health, the two of us would hope to live out the century that is legally permitted. If we live longer than the legal limit, we must appeal to the courts, giving a good argument for this dispensation. I hope that I can set down these thoughts coherently while our health continues to be stable.

Finally, I plan to send a copy of this memoir to the chair of the judicial committee of the Secretariat. It should be seen and read not only as a reflection of the efforts by my colleagues and myself to put into place this new system of life but also to set forth the arguments of the adventurous thinkers and doers who attempted to lift the world community out of its torpor, out of the paralyzing malaise that overcame our planet after the horrors of the preceding century and beyond.

The Secretariat can then store it away, destroy it, even publish it at a time that seems appropriate. Perhaps it will generate further thought, humorous parody, or as I hope, historical perspective.

Chronology of the Revolution

2025–2050—Beginning of fossil energy crisis; attempts to save international economic, political, military alignments, treaties; UN irreconcilable ideological divisions

2050–2070—Struggle to obtain secure fossil energy resources; beginning of emergency national investment in nuclear type electrical power; crisis of the welfare state, fall of dollar and euro, worldwide inflation; new international alignments solidified. International economic failure

c. 2070–2150—Chaos, climate change, war, terrorism, Dark Ages; population destruction

2150–2180—Climate stability begins; emergence and beginning reconstruction, first international medical and scientific/philosophical meetings; power of the intellectual, military elites.

2150—Beginning of voluntary experimental international eugenics, ongoing international gene transfers. Demographic planning
2180–2200—Geneva Directorate of the World Society
2180—Beginning of systematic genetic reconstruction, theory and experiment, Geneva conclaves. Systematic population controls begin.
2200—Transfer of World Society to Nairobi, Kenya/Africa

· 1 ·
A REVOLUTIONARY VISION

Hope

At one of the first general conclaves of scientists, philosophers, and political thinkers in Geneva over one hundred years ago, c. 2180, a scientist/biologist commented that the world community might be on the verge of understanding that we humans are biological creatures. Our culture and history are the products of our evolutionary sojourn amongst the many billions of struggling life forms, which have inhabited our planet for approximately three billion years as living matter has struggled against the entropy of the universe.

Another participant at this conference replied that we are nature's product, with a brain that is both nature's gift and its curse. All sections of humanity were now attempting to map out a direction for the sustainability of the human race on planet Earth.

A number of contributory surface elements precipitated the great collapse of the twenty-first and twenty-second centuries. These precipitators were the issues of demographic indigestion linked to the gradual cloud that "peak oil" ultimately represented, producing economic instability in managing the vast hoard of economically demanding humans. In addition, the intellectual armory which could be mustered to manage or control the ever-more interlocking character of human groups on the planet was severely remiss in its understanding of our evolutionary mission on earth.

Two factors were precipitous in the cloud of incompetence that rendered nil the hopes to resolve systemic failure. First was the diversity in the ability of various populations to respond to modernity and prosperity. They lacked the educational competencies to achieve their goal of modernity and middle-class stability. Second, the constantly recurring ideological and emotional responses to the challenge defeated all our hopes. These elements obscured the vision that humanity required in order to evolve toward a species-wide unity of understanding and direction. The results: the ongoing futility in terms of political decision making. Next were the horrific genocides, war and religious-ideological destruction which catapulted group against group in endless and hopeless bloodletting.

After the catastrophic unraveling and the miasma of hopelessness which followed, there came a flicker of light. The resumption of the old climate sequences helped. An internationalism that factually studied and questioned the old verities wherever humans gathered became a wedge in the door of history that now led to the light. What followed was the need to constantly bring forth ever-more young and rational scientific minds.

These men and women became our touchstone for progress. Their evolving skills, after decades of scientific, technological, political and economic advancement, achieved a self-conscious leadership that was joined by others who saw potential benefits from the new leadership. Gradually decision makers came together to ferret out the old errors and push rationality and secularity to the fore. Our good luck lay in the fact that scientific endeavors were still being pursued.

Organization

We realized that we needed to face up to existing knowledge that had been buried by the ideological romanticism that yearned for unity to be enforced by the gun. This unity could, however, never exist given our incomplete bio-social evolution as a species. Educated communities understood that our species desperately needed professionals in medicine, physics, technology, and a wide variety of hard disciplines to lead us back to reason and competency.

Emotionalism, ideology and religious irrationality had to be abandoned by the vast majority of humans in search of factuality so we could pursue timeworn dreams of peace and prosperity. It was necessary to create a world in which everyone could pursue enlightenment without recourse to ideological fantasies. These visions generated the demonology of hate.

When the fires had finally cooled, the world existed as a virtual police state in which military juntas controlled large national populations. They maintained peace and distributed the vital resources that existed on earth. Despite the vast demographic shrinkage, billions of impoverished humans still subsisted. What was available in terms of natural resources had now to be mined, harvested and turned into food, warmth, and shelter. It took time to teach the masses that the dream of good times preached by the redistributors of the twenty-first century was a mere hallucination.

This military-enforced stability had transitioned into a miasma of hopeless lethargy in the great demographic concentrations throughout the world. But even as humans sought to understand the realities which led to this desert of death, civilizational strivings were maintained. These small oases asked the powers that be to join in the search for long-term solutions to the paradoxes of human life. A few optimists still believed that the best of the past could be embodied on the way to a long-term future.

The Plan

The second half of the twenty-second century saw the confrontation with historical issues discussed with a verve nonexistent in the earlier decades of that century. Some of the earlier communication technology was functioning again. Although actual transport was expensive and difficult to access, people could communicate electronically face to face. It is customary for people to search for *confreres* to meet and plan.

The most interesting part of the new internationalism now brewing was the militarism that had stabilized much of the world and was surprisingly in sync with technologists suddenly sprang from previously hidden repositories of knowledge. This was a happy international awakening from hibernation.

Gun-toting brutality was displaced by hi-tech militarists who quickly shed their uniforms for the whites of the laboratory. By the end of the century long-distance communication had transitioned to meetings in Geneva Canton, old Switzerland. International intellectuals began to hammer out an inclusive plan for all the ethnic and national profiles on the planet:

A. Dissolve the superficial divisions of race, class, ideology/religion that had historically separated humankind into the various war camps.
B. Reduce the population of the world to a level of possible long-term stability and survival given our current knowledge of resources, climate change, ecological richness and scarcity.
C. Husband our energy resources to create an egalitarian access to fossil fuels, and other resources as well as the search for, and the cost effective building up of our renewables
D. Create over time an even playing field of educability in our world population, now to create one panmictic species of *Homo sapiens sapiens*. Hopefully, with generally equable intellectual potential, humans throughout the planet would now be able to 'speak' to each other.
E. Develop over time an international system of governance that can maintain the egalitarian internationalism that uniformly high intelligence and basic social class equality will gain for the species.
F. Maintain a dynamic of material, technological, scientific change that will not burst the dam, here ensuring equable long-term peace and prosperity.
G. Find and exploit the social venues by which the enormous intellectual and emotional energies of humans might be siphoned off, away from war, ideological enthusiasms of millennial salvation into productive 'small-bore' cultural creativity and independence.

Internationalism

The revolutionary significance of these conceptual principles was that they concentrated international overseership in a centralized forum. Material existence was monitored and guided in terms of innovation, social disruptions, and progress.

We needed to control the enthusiasm that had generated vast material, technological and political/economic changes. But such bursts of enthusiasm had too frequently unhinged our understanding of our deeper cultural, ethnic and religious dispositions.

So-called material and technological progress had resulted in a spiral of mental confusion, inevitably causing both fear and power accumulation. Predictably the next phase in this so-called technological progress was the accumulation of national and feudal power that has always precipitated war and genocide. This we voted to stop. Item G of the Geneva plan provided the greatest level of controversy and concern. This complete political control of the material existence of the different nationalities and ethnicities carried the possibility of freezing the natural human creative dynamic. Nobody could know what impact such international control could have on the rationality of the human species and its ability to utilize the functional process of deliberation.

Great fear existed in all participants in these conclaves that such overpowering control of material life might possibly short-circuit the creative juices of humanity in the areas of life design, building, communicating, traveling— personal and social activities that had created the best of civilization— from west to east and north to south. Would such international controls do inexorable harm to the future of humans on Earth?

Humanity had just undergone hundreds of years of uncontrolled and undisciplined material change, vast inequalities of power always with the potential for excising individual self-realization. This uncontrolled chain of events resulted in the despoliation of our planet and the evisceration of our mineral heritage. The greatest catastrophe of all was the loss of our humanity. Our Earth was unable to maintain the civilizational heritage that defined the best of humankind. Humans throughout the world were angry, demanding that they be provided the same amenities as the privileged one percent. The consequent collapse of twenty-first century promises, death by war and terror, the triage of billions of humans, are today understandable. We had created our own millennial disaster.

These questions, raised more than a century ago, still persist and will be front and center of this memoir. I will discuss in greater detail our recent understanding of the historical processes that led to the disasters of the past. This will be followed by a recanting of our contemporary institutionalization to define the reality of our program. We wanted to heal and progress and tried inchoately to follow the message of science that has begun to decipher the mysteries of our nature.

· 2 ·
INTO THE TUNNEL

Tough to Understand

We believe that the past 150 years have constituted a revolution for the human race. This revolution and the goals that we have set are hardly concluded. We have a long way to go before the leadership of international and national governments and among the scientists, technologists, businesses and independent groupings of citizens can declare our 'mission accomplished.' In reality it never will be. As I will discuss, human nature and its future direction and development will always be a mystery.

We do not fully understand the meaning of our evolutionary burst into biological dominance nor the relationship of this evolutionary position in terms of its social meaning. How can we live together in cultural units with symbolic differences and yet maintain a peaceful momentum of change, cultural and social—our plan's item G.

What we have learned so far was, of course, stimulated by the catastrophic social dynamics of approximately three centuries of disaster and progress—the twentieth, twenty-first, and half of the twenty-second century. Our historians have been squabbling, almost without a break, in attempting not only to describe the nature of the events, among which were scientific, technological, medical progress, arts of great depth, continuing worldwide bloodletting, war, terror, genocides seemingly without end. Of course amidst this cascade towards disaster were the ideological pretensions of those in power, their hallucinatory declamations, and the high living that juiced their self-serving spiel.

The outcome was a breakdown of existing worldwide policy resolution mechanisms. So, for over a century we were 'blessed' with triage.

A more concrete delineation of the events and causes of this terrible humanitarian and social disaster requires that we briefly discuss the genesis of the near worldwide ideological insanity which allowed the human species to run off the tracks of bio-social adaptation, peace and progress.

Prelude to the Fall

How do we explain the mess that the human race got into? Our contemporary consensus is that it all started in the 500 years preceding the final 150-year unraveling. The growth in power and demography of the European West started in the days of Christopher Columbus, at the cusp of the sixteenth century. Of course during the previous several centuries the stirring of the European mind in terms of philosophical, mathematical, mercantile, technological invention and progress did provide a base for what followed. The printing press appeared around 1450. Those times were emblematic of the mind's need to express and act on our intrinsic curiosity, experience social change, all energies of the mind that had to be released from binding religious mysticism and control.

There was thus an existing technology for the dissemination of the churning debates about the nature of the universe as well as humanity. Columbus and his followers needed the compass and the astrolabe to help guide them over the oceans. It was a world beginning to move scientifically and technologically. The human social and political revolution began in the seventeenth century. Here this freedom of the mind that curious humans had wrested away from theological supervision was then extended to the political and social realms in what is called the *Enlightenment*.

The assertion of individual rights and the implicit and explicit demand for a democratic polity was in harmony with the expanded movements of mind and body. Humans have historically searched to secure the body to allow the free operation of the mind. We strive to attain a more comfortable material existence and obtain control over our physical environment. Better transport, especially with the invention of the steam engine and the development of coking coal, alerted people to the international possibilities of this economic search for freedom. In the Americas, however, this search for wealth and control involved racial slavery, a perennial moral burden within the trappings of political democracy.

By the end of the eighteenth century, the western mind had been unleashed. *Laissez-faire* economics and politics helped. The scientific method opened up every area of human economic, intellectual and social intercourse to the pulsations of the cognitive mind: industry, agriculture, medicine. And the people flowed out of the hinterlands into the cities to "enjoy" the fruits of the new social life.

Whereas the world population in the days of Columbus might have been in the range of 500 million souls, by the time of the death of Thomas Jefferson and Ludwig van Beethoven, c. 1825–30, it was almost one billion. The inventions poured forth; the fruits of the scientific mind cured disease, increased food supplies, warmed homes. Humans really began to live better even as their numbers were exploding.

March to the Cliff

The nineteenth century produced the capitalistic corporation and the new military and economic power of nations and ethnicities that pulsated behind the scenes. And as always, the old theses bred antitheses, and a new intellectually dynamic synthesis began to weave itself into the external fabric of society and the human mind. This was the inherent social inequality which this new revolutionary form of industrial urban life began to spin off. The planet was getting crowded, but petroleum was now juicing the rubbed motion of all these elbows.

The eras of divine right monarchs and churchly institutions had rationalized human inequality. For a long time this system had quieted down the plebes. The static economics of life, the slow diffusion of ideas, primitive forms of transportation and communications allowed for the acceptance of such inequality. In the nineteenth century this was no longer possible after the invention of the telegraph, the telephone, steamships, and railroads. Our world was moving very rapidly in terms of innovation and social change. The traditional conservative institutions of religion or political authority were teetering at the end of the diving board.

In truth there was a movement throughout the western world to institute educational and other social programs for the masses pouring into the cities.

Science and technology as linked to the industrial system were moving ahead at an amazing pace. Yet the cities, even London and Paris, still stank. The cities yearned for the expansive engineering of classical Rome to flush out the effluent. In retrospect there was little Enlightenment intellectual direction to help us understand these dynamics from an historical perspective. The exception was the socialistic and communistic vision given reality by Karl Marx.

His writings became the superego of the attempt to understand what was happening to an ever-crowded and newly impoverished world. The masses were becoming alienated; the old gods did not explain their plight. The secular intellectuals found an explanatory center in the demonology of class warfare. But who were these capitalist demons? Were they not joined at the hip with the scientists and the inventors who had created this social progress, this enormous wealth?

As we closed the nineteenth century, the population was now 1.8 billion. Few understood the inner human dynamic creating this physical revolution in the character of our social and economic life. What was it in human nature that was roaring past all attempts, with the exception of the flawed visions of the socialists, to understand the depths of human creativity and power? One thing was for sure, it first had to destroy the old order and then inflict upon mankind the new, a unique inspiration of bloody conflict.

In reality the twentieth century began with the Great War of 1914–1918 accompanied by the first of many genocides inflicted upon innocent "demons" who had supposedly caused the multiple inequalities and misunderstandings that the human race was undergoing. The Armenians—a minority in a dissolving Ottoman Empire—were the first to experience this inchoate hatred. Once they had been a tolerated, even a modernizing minority group. Now they were singled out as Christian and middle class, 1.5 million innocent subjects to become the focus of genocidal vengeance among the Muslim majority.

Europe and America flexed their new musculature of industrial power and political ambition. In the end WWI saw the final dissolution of monarchical pretensions, which was a prelude to something even worse—totalitarianism. Millions of young men died from 1914 through 1918, their graves a bridge toward an even greater world conflagration in WWII. This time death was undergirded by hallucinatory ideological persuasions of the masses, an insanity that eliminated tens of millions more.

The Plunge Begins

The League of Nations lasted for about twenty years. Founded at the end of WWI to bring the nations of the world together in peace, it was set adrift by the beginnings of the next worldwide conflagration, and few nations missed its presence. WWII was in the making, now rationalized by salvational secular ideologies—socialism, communism, fascism, national socialism, even welfare democracy.

We need not traverse in detail this bloody ground which has had so many expositors. In the buildup to WWII there were the 1930s Soviet communist genocides, in which they slaughtered their own citizens—the supposed capitalist exploiters and their intellectual kin. The epochal (*Holocaust*) occurred during the inferno of WWII, the planned industrialized extermination of 6 million European Jews by the German National Socialists and their European-wide *confreres*.

Europe lost its finest young men, those whose fathers had seen that earlier butchering of the virile and talented. That loss was greatly extended to other populations of innocents. The Japanese in the East rampaged through Asia. The miracle was that this nation and Germany subsequently returned to the family of advanced and wealthy democratic nations. Victorious Europe had now lost its historic élan. A classical scholar, Gilbert Murray, has described the fall from greatness of Athens. Once the glory of Greece, now defeated and humiliated, war-weary Athenians witnessed their young women playing the flute and lyre to accompany the Spartans as they supervised the throwing down of the long defensive walls around Athens: The Athenians, subsequent to their blood letting: '…lost their nerve.'

The end of WWII saw the United Nations established in the city of New York in what was then the United States. It endured fitfully for less than one hundred years, with great pain, it should be added. For half a century following the WWII conflagration we witnessed an odd time in human history. During these few decades the chastened West and East turned their backs on war as a pathway to prosperity and power and utilized the industrial and technological knowledge now accumulating to create a momentary prosperity for their citizens. Petroleum was plentiful and cheap; the banks were willing to lend out a river of debt, and the socialist-redistributionist model gave an altruistic cast to these policy endeavors. It was a time of political fat cats as well as corporate elites.

The new ideological vision, humanity's superego of moral policy making, was equality of condition, rather than the now-ancient Enlightenment vision of equality of opportunity. Here was the Marxist apotheosis redeemed in democratic-socialist garments. The sudden awakening of world consciousness to the destitute colonials of the planet and the so-called Third World citizens and migrants within these now-ebullient nations led them to funnel enormous amounts of welfare assistance to the many new nations that were mired in poverty.

During these decades the oil gushed forth from the earth to fund this morality play. Into the Third World the monies flowed, sadly down the drain but thence, miraculously into the pockets of indigenous thieves who ruled over these hapless peoples. Still residents of the Third World responded to the wealth and the medical 'death control' that had been visited upon them by the good-hearted folks of the First World who had somehow forgotten that 'death control' in the developed nations always needed to be linked to birth control. If not, the imbalance would surely create a Malthusian disaster. By the year 2000 the population of the world had broken through the membrane of serviceability to reach 6.2 billion from the 1.8 billion one hundred years earlier.

Recall that several hundred million of the best-educated people had perished by wars and genocides in the interim. The wars and genocides continued after 1945. The United States, then the greatest power of the twentieth century, along with its so-called democratic allies, engaged in wars in Korea, Viet Nam, Afghanistan, Iraq, Syria, with these conflicts leaching into the twenty-first century. Vast genocides continued in tandem. First in China, then in Cambodia, all instigated by communist rulers against their own people. Bangladesh and Indonesia experienced widespread killings. Nigeria, Rwanda, Sudan experienced a great slaughter of innocents. Europe itself saw the Muslim converts of once-Ottoman Bosnia suffer the fatal weight of Christian hatred.

This awful century, the twentieth, left us with sharp contrasts throughout the world. There was great wealth in Western Europe and North America. The Far East was on the rise; China especially began to dominate under its autocratic state capitalist economy. Most of the Third World required the Western dole. Islamic terrorism was on the rise: the small oases of oil wealth, again controlled by a few super-rich autocrats, were surrounded by masses of scientifically ignorant Muslims yearning for the dreams of ancient glories. They did not at first hate their co-religionist exploiters, but the Jews and Christians down the road, who were using their secular intelligence and education to modernize and prosper. The hatred and terror were not successful enough with the Others, so they turned on each other, Sunnis contra Shiites, all under the supervision of Allah.

How do we look back upon this era? Our conclusion: Intellectually and philosophically, no one assemblage of thinkers throughout the twentieth century truly 'ran the ship.' Deep, dark ignorance prevailed. Significant scientific knowledge existed that there was a transitional species undergoing in parts, enormous material, scientific and technological change. However, the searchlight was not on the deeper intellectual and emotional factors that spurred this advance.

Rather the concern was with surface manifestations, the effects of this great social shift on material features, economic outcomes, not causes. The chaos moved into the twenty-first century; demography continued to swell. The leadership groups were unable to understand and manage the chaos. Ineptitude was joined to the hip of personal venality. The outcome of these events could have been predicted from 'up on high.'

· 3 ·
COMING APART

Phase One

We focus our lens, if briefly, on the first half of the twenty-first century. What seemed to define the perpetual crises of this period was the lack of intellectual and moral directivity. It was as if the leadership were wandering in a mental fog about causes and solutions. It was always patching things up, retaining the same institutional arrangements and power structures. What seemed to have worked in a time of growth, the sixteenth to the early twentieth century, did not work anymore. The rhetoric of equality, democracy, free enterprise, socialist redistribution bounced back into universally deaf ears. It seemed that the elite leadership was marking time to fill their pockets and those of their own people, hoping to be underground when the tornado swept down upon the innocents.

The first decade saw the collapse of European socialist debt economies. The United States was in somewhat better condition in this decade because of its own oil, natural gas and coal reserves and somewhat more open economic structure, but spent its reserve currency on debt and the redistribution of this debt to its growing 'indentured' population. This economic situation, however, did lead to an energy 'breather.'

By the third decade oil supplies plateaued and then diminished; prices rose. Natural gas was in somewhat better supply but also rose in cost and became scarce. Renewables, because of their lack of scale in the great urban complexes, became objects of conspicuous class dalliance for the wealthy and sun drenched. Because the rise of the oil economy had led to the enormous growth in wealth throughout the West in spite of the recurring savagery of wars and genocides, it is understandable how a shortage of this vital lubricant would shift the political alignments.

Small-scale wars had continued throughout these decades; terrorism spread from the Islamic world to Europe and Asia. The advanced nations deprived of energy and raw materials made a rush into Africa. This continent was definitely on the road to becoming multiracial and multicultural. Large numbers of Asians and Europeans were now in military residence, each carving out their neo-colonialist economic interests and hoping to batten up the flagging home fronts. Naturally the indigenous Africans received their colonial pittance.

The debt bomb of the 2020s unraveled both Europe and the Americas. The euro disintegrated, along with its political bonds. In the United States the movement for the disestablishment of centralized control gained ground as the central government could no longer fund a dependent population that had been created by America's subsidization of poverty. In the anarchy of scarcity, taxes dried up; the currency dissolved; inflation tore through the national fabric. Those few states in the U.S. that had self-sustaining productive economies wanted out. There was great political and intellectual conflict as this once-great nation lost its affluence and influence.

While the old West fragmented and retrenched, a new center of political, military and economic power arose in the East. The old animosities and fears faded as Russia, Japan, the Chinas and Koreas formed alliances. Underpopulated Russia and Central Asia would cede control over their vast natural resource base to the Asiatic industrial goliaths in return for their manufacture and protective military cloak.

These Asiatic nations, then ethnically homogeneous, would unite into a politico-military-economic sphere of influence and bring the rest of Asia within their power embrace. Africa and the Middle East were drawn into this new alliance powered by its scientific, engineering and technological prowess. Those were nations in which the citizenry were still willing to work.

In truth this political/ economic/military structure held up only as long as the spigots of oil and natural gas continued to dribble, 2020–2035.

The Crunch

Mid-century however, was the pivotal moment in the drying-up of the flow of black gold. Several of the remaining large oil and gas fields had already failed. The now-scarce heavy oil of Canada and Venezuela required increased refining and consumer costs. There was a concomitant splurge of scarcity investment throughout the developed nations to rehabilitate old uranium atomic energy facilities or build new ones for thorium and with other new energy sources. Research into fusion processes was initiated by China and cooperatively by a number of the seriously weakened European states.

The theme was, of course, electricity, the foundation of our international civilization. Without the energy needed to produce electricity there was to be no national survival. Fortunately the competition for these new mineral resources did not result in new wars. Most people were already poor by mid-century. Now there was almost universal impoverization. The once-dominant and now barely surviving nations began to make frantic (desperate, if coordinated) political/economic agreements. As the decades began to slip by, those without power, resources or human energy began to be subject to the triage of 'being ignored.' There were no atomic wars for dominance but much internal disintegration, lawlessness, disease, pandemics, and never a surcease of random terror.

It was clear as the world progressed ever deeper into the twenty-first century that the dream of the United Nations to invoke a rational unification of humankind through international law was proving to be a chimera. The *weltanschauungs* of the various national leaderships could not come to grips with a mutual diagnosis of the anarchy taking place around them. At some point during these mid-century decades, the world population passed the 8 billion mark, reaching a peak somewhat north of this figure around 2070.

The old ideologies still 'hung around.' The power structure in place was still attempting to plaster up the cracks. But those repeated late twenty-first century patterns of the 'same old' new elites were showing their age. A new reality was inexorably seeping, creeping into the awareness of those blessed to survive. *Smart people had all the power—those who kept the lights on, scientists of all disciplines, technologists, healthcare professionals, the military and the police.* Old expectations of middle-class life had evaporated in the agony of scarcity. Force was ubiquitous, mobilized by the skilled to protect their own privilege from the mobocracy boiling up out of those billions who were being passed by.

The constituent states of the United States along with an ethnically fragmented Europe, Russia included, were drawn in to mobilize their once-great military establishment to maintain the peace internally and internationally. The cries from the masses (those who could still be 'heard') were for an autocratic economic egalitarianism. They had been taught that social inequality was artificial, superimposed by *Others*, certainly not by those of their own class who had profited from their political power. Supposedly it was the role of government to equalize the condition of *all*. These still-functioning nations were caught in a dilemma. Not having the taxable wherewithal much less anything distributable, they opted to use the police/military establishment to maintain what productive capacity there existed for the base survival of any who could survive. Centuries-old liberal redistributionist commitments evaporated in the overall anguish. Their own jeopardy and survival decided the internal debate.

It is no accident that the events of the mid-twenty-first century and beyond sound like the unraveling of ancient Rome. The late twentieth century conditions for prosperity—plenty of cheap and fluid energy, more resources than people, and banks full of loans—had evaporated. No more rabbits could be pulled out of this political hat. Unity of the world, the UN, great nations and alliances were *poof*. Everyone wanted out. Secession, states' rights, independence were the catchwords of the late twenty-first century. St. Augustine (fifth century) would have had his heart warmed by the dissolution of the secular monolith.

Final Blow

The chronicles and histories of that era give one a sense of the crashing of a tsunami on the unsuspecting citizens of the world. A dark cloud, once barely visible on the horizon, of a sudden now cast its shadow over their lands. Saudi Arabian princes had to once and for all end the gasoline and oil subsidies for their 40 million desert inhabitants for the sake of purchasing grains from foreign sources and subsidizing the building of salt water purification facilities. At least that was the propaganda of the princelings. The bloody riots shut down the kingdom for many years. Chinese and English troops and warplanes protected the depleted oil facilities.

The same events occurred in the Gulf Emirates as they deported the millions of non-citizens who labored for them. An Arab satirist, luckily far from home, created the following homily:

> The great grandfather rode his camel with dignity; the grandfather's chauffeur drove his Buick to the oilfields; the father had two pilots guide his jet around the world for him to inspect his real estate holdings; the son drove his own Camry to the family watering hole; the grandson had a motor scooter to get to his doorman's job; the great grandson spent his 'inheritance' in transforming the family garage into a barn, to accommodate his camel.

For most of the people of our planet in those final decades of the twenty-first century it was not a time for laughter. The Russians were scouring the Arctic and the wilds of Siberia with their Asiatic allies, searching for oil. They would finally acknowledge that the West Siberian fields were permanently dry. The Americans were no longer subsidizing unwed mothers and their progeny; the available monies now had to provide gasoline for the police and the army. In Europe, the 35-hour week was more like a half shift of 20 hours (when even that was available). Wages were also cut in half.

What once had been a proud political/economic state program for the rivers of redistributionist wealth became dribbles of life support. In the undeveloped nations and protectorates, after the mercenaries had crushed the populist terrorists, an occasional mule-drawn vehicle would arrive to toss out short rations of food and water to the apathetic and hapless of Africa, South Asia and Latin America.

It is a tired cliché to speak of 'the straw that broke the camel's back.' But the truth is that events after 2070 sent the entire world into a spiral of dissolution that lasted for over half a century. By that date a new leadership generation was searching for the light of hope.

Climate change had always been a theme of the "liberal" sage who saw private-sector industrialism and technology as a threat to his/her dream of universal equality. It never bothered these ideological 'visionaries' that government control of technology and industry in the Stalinist or Maoist incarnations always led to downward equality and the opposite of what they had been campaigning for.

The weather did change in those final twenty-first century decades. The scientists hypothesized that the interstadial (which had allowed innovations in agriculture that, in turn, allowed urban civilization) that had superseded the last Ice Age 12,000–15,000 years earlier had possibly ended. No rain came, only a snowless cold. The monsoons became spritzs; vast swaths of the southern continents were desertified. Macro-tragedy struck. Those political alliances of scarcity disappeared; the world went feral. Defense of the border and internal lawful behavior became humanity's primary vocation after the search for electricity, warmth and food.

In the United States, the former United Nations compound was now rented to the squires of Latin America; they jumped ship recognizing this new American confederation's militarized quiescence. Naturally, they were able to transition to New York City with hidden gold and diamonds, claiming a condominium here, a penthouse there, always armed guards for themselves, now a signature emblem of a perverted prosperity.

In Minnesota, USA, the chronicles tell us of a mother speaking at a mass meeting of the citizens with the leaders in the capital city who were considering sending food to the starving masses of desertified Sao Paulo, Brazil.

> We have a son; we are a family of three. Both of us slave away to afford wood and coal for heating. At night I stand in line for hours to exchange my food stamps for bread, all to feed, clothe, and educate our son. You want to send our food and precious water to Sao Paulo in Brazil to feed women with their four or five children. Yes, they are hungry. But why didn't they think before they had all those kids?

These are harsh thoughts for those of us who today live in peace and adequacy of our twenty-third century. That Minnesotan's harsh thoughts and words would have pained her ancestors in the halcyon late twentieth-century United States.

There were attempts to help. India suffering horribly from starvation and social chaos received some petroleum and food for their elites from Russia, now a political ally. China and Japan sent engineers and elite troops to Ethiopia and Sudan to help turn back the remnant of the Egyptian army which had tried to destroy their dams. The lower Nile had dried up and its 100 million citizens now were starving. In antiquity when there were a mere 5 million Egyptians, its annual harvest of grain fed Rome for four months. Egypt itself basked in Roman peace and prosperity. By the end of the twenty-first century it was disappearing as a nation and as an ethnicity.

Even the United States, once the magnet of the poor of the world, especially from 'south of the border,' had to use its waning resources, to defend this border. The state of Southern California, itself with a majority Spanish speakers, joined with other southern border states to deter the desperate from the south and deport the non-citizens from within. To the east the Rio Grande River had become a dried-out arroyo, a cactus-strewn walkway now ringed by barbed wire and mines.

Hope

From our standpoint 200 years later, there was only one bright spot in the late twenty-first century. It hinted at a future for this ever-more confused and predatory world community. About the middle of the twenty-first century, the Chinese had begun to perceive a trend. Their population, which had reached over 1.5 billion within their continental borders, began to give the leadership nightmarish visions of their future. Still an autocracy with functioning operational national politico/military controls, they began to institute a series of lotteries and awards to allow certain segments of their populace to have one or two children. At the same time they put into place severe restrictions on child bearing by the masses.

The Chinese who paid significant taxes were given bonus points, and the bureaucrats and military were likewise favored. Scientists and important cultural figures were also granted the privilege of joining this new 'mandarin elite' now mandated to reproduce in a modest way. The lottery chits were restricted to those considered to be the most highly valued bio-social segments of the society.

In the West of the same period, the demographic flood of poverty was challenged by a wild mixture of individual survival skills. It was everyone for himself. The demographic issue was to be dealt with by nature. From a distance they interpreted the Chinese solution as typically characteristic of Chinese autocratic polity. The rulings in China were to the morality of the West, a tyrannical invasion of basic human rights. Westerners considered it their right to have as many children as they could afford. Ironically, only a few decades earlier, so-called liberals argued that women could bear as many children as their neighbors (society) could subsidize.

Quietly other Chinese communities as well as their allies in the Japanese islands and the Koreas began to follow suit. Several Vietnamese provinces and Thai communities took up this new idea. These nationalities were fragmenting by the mid- to late twenty-first century. Wherever there was ethnic unity and a strong military/political structure the Chinese experiments in survival were being put into place. That is, they were if the powers that be were able to neutralize the intermittent clarion calls of racial or ethnic discrimination as to whose genes should be passed forward on. A few enlightened leaders attempted to look forward toward survival. At the same time they may have been looking back at humanity's heritage and saying, 'We must now gain control of our destiny.'

· 4 ·
LIGHT AT THE END

Dark Age Demographics

The so-called Dark Ages were said to have begun with the occupation of Rome in 410 by the Christian Visigoth Alaric. Then a bit more than a century later the occupation of Ravenna, the last redoubt of the Roman emperors, now Christians of one sort or another, by the Ostrogoth, Theodoric, marked the real beginning of centuries of darkness and minimal cultural and social advance in non-Byzantine Europe.

The late twenty-first and twenty-second centuries of surrender and loss of hope were relatively brief moments in human history in the long run. Signs that nations and their leadership understood the nature of the ravine into which most of the world had fallen into began to surface. It took considerable time to gather together a majority of nations, peoples, intellectuals to reach a consensus of causes and diagnoses in the hope of fashioning a plan of action to once more restore a semblance of civilized life and law.

Perhaps the demographic calamity that was then in process of wiping out much of humanity helped turn the collective human mind toward factual reality. Demographers in the early twenty-first century focused on the amazing ballooning of humans on this planet—from c. 1.8 billion about the year

1900 to 6.2 billion in 2000—but avoided issues of cause and solution. They used their computer models of extrapolation and concluded that the population of the world would be about 9 billion to 9.5 billion by 2050, perhaps as much as 10 to 12 billion by the end of the twenty-first century.

Then, the sociologists hypothesized, the world population over the next century or two, would level off and begin to gradually fall. They assumed that the Third World poor would soon follow Europe, the United States, and one hoped, China and India, in also becoming middle class. To preserve their new-found prosperity they would use modern medical technologies to limit the number of children born. Of course all these hypothetical models were ideologically driven by the powers that be, who were enjoying the fruits of mass production and the wealth that was falling into their hands through this population explosion.

The reality was, of course, different. As the twenty-first century ended in chaos and pain, demographers were estimating that the population had actually fallen below 8 billion people, the number of humans estimated to have lived on this Earth in 2025 CE. The decline continued into the twenty-second century; the essentials of life continued to be scarce, the unfortunate climate change was still unrelenting, and there was ongoing tumult and disease.

And of course, as we have seen in the roughly 200 years since 2100, there are far fewer humans on this planet as I write. The mission to balance and stabilize the future of this species is still incomplete. I believe that the leadership will remain firm in reducing the press that this human weight exerts, allowing humankind to maintain that extra savings account for programmatic flexibility that could meet nature's inevitable environmental surprises.

Triage

Our historians and social scientists in those parts of the world which could afford historians and social scientists over the past several centuries have gone wild in their attempts to both chronicle and understand this descent into madness and death that the world experienced over the course of that cataclysmic century. The worst of it lasted from the final decades of the twenty-first through the first quarter of the twenty-second century.

Our young people have wondered as they have studied their texts how such events could have occurred given the civilized status of humankind, the enormous amounts of knowledge stored in libraries and electronically communicated daily to the denizens of this planet. Certainly this was true at the end of the twentieth century and well into the twenty-first.

The facts that stand out in my memory during my own youthful studies involve the horrible events that occurred on the various continents. The Canadian populace was at one time quite welcoming about their foreign immigrants. At the close of the twenty-first century, however, they were separating men and women migrants from the states below their borders who had avoided the mounted police and military blockades at the borders. These humans were products and then victims of a welfare system that was no more. They were now hungry refugees. Young children stayed with their mothers. The old were allowed to remain together. Why the separation of men and women? The Canadians wanted no more children of the desperate poor, the losers in the Darwinian struggle. If the refugees agreed to sterility therapy, the Canadian refugee authorities allowed them to unite with spouses and be placed in camps.

Events during the next century, the twenty-second, in the Middle East, Latin America, and in Africa were far more horrible. What once existed in a fragile political and economic environment had virtually everywhere (except in South Africa) dissolved. Both the residual oil and tropical rains and water sources had evaporated. In Africa the continent, from the shores of the Mediterranean south and east, was being overrun by terrorist gangs. There was no organized indigenous political institution to counter the pillaging of roaming extorters and killers. A military force was organized through special contributions by nations that were poor but yet functioning, both in Europe and in East Asia. Armed with existing modern technologies and armaments they were equal to what these gangs could muster, the gangs were hunted down year after year, given no quarter. The ominous phrase, 'no prisoners' was the rule.

As it was, the middle class from all the afflicted continents, from what were once wealthy nations with natural resources and industrial clout but now largely barren wastelands, was being picked apart by hardly less nourished vandals. Populations were disappearing into the mists of carnage and emptied cities. In South and Central America, wealthy militaristic autocracies were being reincarnated by the very rich elites, many of European ethnicity. If there was starvation, so what?: The army was paid; the walls around the vast *favelas* high and covered with electric wire. If the people wanted a piece of bread they had to subject themselves, like the refugees from the south into Canada, to sterilization.

Over the decades in the twenty-second century, the violence abated; the poor and vulnerable disappeared. Aside from violent Africa and the Muslim Middle East, population reduction of humans throughout the globe was now proceeding under that silent and misty cloud of supposed ignorance that enveloped the world during the (Holocaust), 1939–1945. Worldwide communication and travel almost disappeared, who could pay for that? No information, no leisure, no moral outrage. From the standpoint of the developed world of the early twenty-first century no one could have imagined that triage on such a vast worldwide scale could occur. We had to educate our young to the realities of the period 1939–1945.

Australia is an example of the survival of a few. The continent had become a great desert except for a small southeastern enclave where the city of Sydney still survives. There were a few outliers to the worldwide darkness. Scientists in distantly removed geographies, such as Sweden and New Zealand toward the end of the twenty-first century were able to make significant progress in turning the modest technological advances heretofore achieved in hydrogen fuel-celled energy production. Here began a theoretical/experimental program to harness the tantalizing profusion of energy once promised by the fusion power of the sun and the hydrogen bomb.

At that moment in time no social impetus or sufficient wealth existed to put into place the large-scale and expensive experiments that might then have turned a world from energy scarcity to plenty. Yet there still remained in the minds of talented humans the dream and concrete vision of a plan to generate from non-radiating atomic energy the electricity for water desalinization, agriculture, medicine, transport, heat and light to advance progress worldwide. This dream existed here and there, even during an epoch when our world was imploding.

Light and Hope

What could be the destiny of a world living close to the feral line without water for crops, plentiful energy sources for fertilizers, pesticides, fuel or abundant electricity for transport and urban life? After a half century of radical climate change and perpetual cold and draught, northern forests were cut down for fuel over decades in which billions of humans disappeared. In the chaos, law, justice, and hope evaporated, and throughout the world people inverted their life dynamics into survival dormancy.

Needing only just enough to keep them alive they lived in the bizarre anticipation of a slight warming which told them of the coming of the summer solstice. Imitating the life cycle of bears in the winter, billions barely moved out of what shelter they could maintain, hoping against hope that life could renew itself. Oh, if only the tales and memories that the elders dreamed about could come true.

A small miracle happened toward mid-twenty-second century, an easing of the bitterness and a few sprinkles of rain. This was followed by a gradual, decade by decade, return to the older interstadial seasonal temperature variances. In those parts of the world where desperation was still linked to strategies for rational survival, intelligent political leadership and police still functioned. These quadrants had maintained the business of civilization, a structure of education, even the skeletal preservation of the arts. Here, with the warming and the water, the people quickly began to pull themselves out of their apathy and pessimism.

In South America the great jungles had been turned into prairies, the cities into ghost towns. But wild edible animals still roamed through the stunted grasses. As the rains appeared, the hunt was now on. It paid to be able to fabricate a bow and arrow, even while waiting for the gunsmiths to conjure up the iron to be made into steel and the lead for bullets. Civil society by force of necessity began to break the chains of autocracy. There were many fewer humans to be managed politically.

In the Northern Hemisphere where the forests had long disappeared, seeds were being planted for crops, trees grafted. The natural innovative entrepreneurial qualities of the human mind were recreating their economies, and the trees began to give shade. The financial structure first put into place in this rebuilding was historical and perennial, barter or rough coinage of silver and gold when it could be mined or retrieved.

There were fewer students to be educated, yet still too many for a surplus economy to exist. Plenty of schools and college buildings needed to be refurbished, even those which had not been vandalized. Teachers had to be educated and libraries reconstituted—so many books had been burned for heat. In certain luckier nations, a bit of oil was found. The fish ran thick in the rivers and oceans now. In a number of nationalities the thrust to know almost reincarnated science and research, even a violin or two for the dance.

As in the ancient post-Ice Age world, agriculture rebounded. In the tropics even a monsoon showed up. The word 'beginning' now meant several years. These moments gave support to the almost lifeless farming traditions.

Doctors and research scientists seemed to appear out of the darkness, a welcome dawn. Progress, albeit first at a snail's pace, began to be a dreamed-of word and soon passionately worked for.

Crossing the Border

During the middle of the twenty-second century we began to feel the collective mind hearkening to new energies of hope that the human race might flirt again with a vision of law, modernity, and the ability to renew the scientific advance.

I have mentioned how the walls of civilization were crumbling at the end of the twenty-first century almost without fail. There now came to be a unity of purpose between the military/police elements in each society and the business/technological leadership, an attempt to rebuild and maintain the fabric of civilized life. The absence of criminal behavior was key to any community's survival. This included the protection of the borderlands.

In many of the poorest nations, there was enough visionary leadership among both indigenous and resident colonials to muster the home guard. And as this leadership oversaw year by year the wreckage and dysfunction, they latched on to a new clue for survival. Stories began floating around of military dictators in Latin America taking steps to ensure the sustainability of their autocracies, so too were the vigorous Canadian and Chinese attempts to sustain the community by enhancing the intellectual capital of each society. In the beginning small tokens of bribery persuaded the most able young women in the society to allow insemination, artificial or natural, by the most talented and able males. The results began to stimulate emulation throughout the world. The African leadership soon was in the forefront of this new vision of modernity. That old political shibboleth of racial or ethnic exclusivity had disappeared in the mid-twenty-second century, and a new political theme began to be heard, 'brain power means survival.'

It was not too long before methods of preserving and transporting male genetic material would allow for the selection and distribution of what was colloquially called 'genius juice' by this new military/political leadership. 'How do we protect our turf?' was the question. The answer, '…smart people creating complicated technologies.'

For the record I should state that it is my belief that one of my female ancestors was an early participant in this genetic integration of humanity. She probably was a daughter of a West African tribal dignitary, a survivor of those terrible times. She must have agreed to this experiment, promoted apparently in a number of still existing West African communities.

This impetus towards racial hybridization was in reality not too different than what had happened during the slave trade of the eighteenth and nineteenth centuries or even in prehistory when the genes of the Cro-Magnon pushed the human species into sapiency. But at this point in time it was completely voluntary and done at the instigation of the black African leadership. The sperm donor, of course, was anonymous, often of Europoid or African leadership origins. The female child of this 'union' was eventually legally married to a male, himself of unknown hybrid origins. And so it went according to what I have learned of the process.

· 5 ·

RECONSTRUCTION

How Many People?

It took several decades for the research and dialectical capabilities of the new generation of thinkers to begin to reconfigure our knowledge. One of the crucial issues was the demographic one. Well into the beginning of this new century, the twenty-third, demographers had gathered enough data to estimate the world population at somewhat over 5 billion. However reports still showed that the efforts towards population reduction were continuing except in those several continental areas that had suffered the most devastating declines.

The Chinese lottery approach seemed to have been one of the more philanthropic methods of dealing with populations which were still large enough to place a heavy burden on the efforts of nations and peoples attempting to regain a measure of economic and social viability. China had, c. 2210, reduced its population to less than half its previous high of 1.5 billion people, even given that it had expanded its borders to what it felt rightfully entitled to as the central nation of the world community. It had held together relatively peacefully during the horrors. At the same time, ironically, the provinces had moved away from the center, revealing the ancient tendency in China, as elsewhere, towards linguistic and cultural diversity. One Chinese estimate even put their continental population at mid-century, 2250, as low as 450 million people.

The international effort during the late twenty-second, early twenty-third centuries was educational rather than viciously authoritarian or murderously tyrannical. The word that issued from the new generations was sustainability. But now it was not aimed at the chimerical vision of alternative energies such as solar and wind. Nor were there discussions mirroring the gathering cloud of fear which 'peak oil' had augured in the early twenty-first century. Humans throughout the planet from the mid-twenty-second century were realizing that their still vast populations on earth were drowning civilization, causing bleak scarcity and denial. Having seen the abuse of this gift of intelligence that nature had provided for humankind over many hundreds of millions of years, they were whispering, *no more*.

Sustainability had to mean that we ourselves had to rationally and universally decide on the balance between the number of humans needed to construct an international civilization of independent middle-class political and ethnic entities. But we also had to think about the need to protect our resources as insurance against the inevitable curve ball that nature could and would throw at our species. And this approach for a new mission on earth had to have a time line in centuries if not millennia.

Specifically we would have to input into our computers, which were coming alive again, what range of human numbers could live on this planet considering the 'black swan' events that could challenge our vision of the middle-class life at any moment in our technological and scientifically created life style. Middle-class security had to mean more than standards of living. It had to take account the diversities in life choices and culture necessary for an international civilization that would not mimic a gray prison warren.

Intellectual Capital

Nations were now facing another basic issue: human capital. As the world fell into disarray large percentages of the populations of many national and ethnic entities were incapable of rising to the challenge that nature and humankind had flung at them. They had been submerged for too long on the receiving end of governmental support, the 'dole' as many had characterized it. Science and the technological fruits of science that were increasingly abstract in nature required an education that was increasingly symbolic and beyond the capabilities of vast numbers of humans on this planet.

In addition to being displaced in an ever-more challenging labor environment, they could not comprehend or intellectualize what was happening around them. Nor could they rise up to participate in the many personal or community sacrifices and energies required of them in an increasingly savage environment. Bluntly speaking, the brain power was not there. The world had inexorably changed. Humanity could not go back to the hunting and gathering economy, the way of life of the primordial eras.

The leadership of the late twenty-second century was no longer the strong-armed militarism of the dark decades. The world was again desperate for the scientific, productive, and creative mentalities necessary to build a new civilization of peace and prosperity. The new intellectual and moral/political leadership that was assuming the responsibilities of power and reconstruction was asking this critical question: 'What can you contribute to your society in terms of ensuring its capacity to defend itself economically, socially, culturally?'

It was clear quite early in the educational and vocational experiences of the young that indolence and moral laxity would work against any particular individual, male or female, obtaining the privilege of a lottery ticket for family formation in any modernizing community. To paraphrase a Supreme Court Justice of the United States, many centuries ago, one generation of incompetents is more than enough of a gift to fellow citizens.

It was agreed that individuals who could not pass muster in this era of the fragile rejuvenation of society would be taken care of in their lifetime. They could marry, work at their skill level, but would be unable to propagate the next generation. The consensus was that individual communities would make these decisions and allow for individual variability, late bloomers, for example. Because tragedy still surrounded humankind in this early phase of renewal, there were always children available to be adopted into solid home environments.

World Society Convened

By the late twenty-second century there was enough social and economic progress and now international agreement that a new world organization should be formally established. The city of Geneva in one of the cantons of the multi-ethnic nation of Switzerland was chosen, as it contained many remaining edifices of ancient international organizations. The World Society was tentatively convened; a Secretariat was established as an international research bureaucracy to represent and administer to the Congress of nations and their peoples.

Naturally, democratic elections were held and representatives sent from a large minority of the various communities, nationalities, ethnicities that had been formed after the various disasters, wars, climate changes, disease/pandemics had crushed the old system of the early twenty-first century. It seemed as if each year brought a few more delegates to Geneva mostly to discuss and plan. The resources for a truly international democratically representative world government were yet in the future

It was decided that in the year 2200, the World Society would move its home base to Nairobi in Africa. All national interests agreed that this continent had suffered so greatly during the crisis and was seriously underpopulated as a result of tragic events. The vision was for a new world organization of peoples and ideas to be here established. This institution would arise from this now ever-more benign ecology and climate. Nairobi would represent a new start for the human race.

Destiny and Human Nature

The great intellectual passion from the latter half of the twenty-second century was analysis: Why had this happened? Historians alerted us to events several thousand years earlier when another civilization had collapsed into war and barbarism. Then people had aspired to a new way of life that would both counter and pass into history the old corruptions, violations of humanity's moral vision. A commitment constructed from a deep sense of urgency turned Saint Augustine toward his introspective vision of Christianity. Augustine, in his various writings about his own pursuit for human reclamation, represented a great historical movement.

In the decades and centuries that followed, the teachings of Augustine, Jerome, Ambrose and many other Christian thinkers posited a philosophical/religious vision that was taken up as representing a much higher and more abstract intellectual and moral sense of the destiny of Rome, now to be 'A City of God.' From our own perspective we may chuckle at the mysticism, at the gestating autocracy that this church at Rome represented to mankind. But, in breaking the model of a degenerate military autocracy, by refilling the withered veins of this ancient empire with new blood, it poured the foundational cement for future emancipations of the mind and the regeneration of science.

Thinkers in each successive decade of our time came up with successively innovative visions of a world transitioning from the old ideologies and their resulting chaos, brutality, and irrationality. That is why all of our mental energies became increasingly focused on diagnosis and prognosis, the great theme of our time: What is human nature? Where is it destined to thrive on this planet?

The analyses seem to have broken off into two directions. One focused on the enormous bubble of human populations from both the developed and supposedly undeveloped corners of the planet. How and why did the leadership groups not foresee that such an expansion could not go on forever? Whose interests were benefited by this expansion of goods, cities and humans that could cloud the minds of our independent intellectual leadership?

Where in heaven were our universities? Were they all absorbed into ideology? Many were now burnt-out shells, repaid with destruction by the masses for the same venalities of power and wealth into which most of the great corporate institutions of culture were suckered into. Like lemmings they had followed the path of bloated governmental bureaucracies. Think of that millennial failure, the United Nations. Here and there, there had been some momentary alleviations. For the most part, it was a castrated institution, permanently riven by ideological incompatibilities.

Where were the cries that with a world population of c. 7.5 billion and growing that there was nothing that could be done, given our natural resource conditions toraise up and maintain even a small proportion of this population on a modest middle-class even keel? And yet the shouts were still overwhelmingly for the redistributing of the wealth from that small fraction of the world that was living high and mighty. Why birth limits? Who will buy our junk and make us rich?

The second focus of analysis was oriented toward a more 'socio-biological' (a momentarily ubiquitous term in the early twenty-first century) analysis of the human species. Did the brutality and impersonal cruelty of events caused by the forced reduction of world populations during the late twenty-first to the mid-twenty-second century (c. 2070–2150), mean that human intelligence itself had failed to meet the challenges of nature? Were the irrationality and ideological fanaticisms of the earlier centuries a cause of those events??

The ancient ideological expletives, 'capitalism, socialism, jihad, racism, religion' continued to linger in the minds of all who shared responsibility for the disaster of the late twenty-first century. The fight against the supposed venality of the wealthy during the nineteenth and twentieth centuries led the masses to follow these pied pipers of totalitarianism into the most horrible genocides of human history. The following century of supposed prosperity supported by billions of barrels of petroleum poured down a sinkhole of conspicuous material consumption continued the nineteenth century vector despite the egalitarian ideological smokescreen created by the power brokers.

But underlining this ugly wastefulness within a culture that was deprived of intellectually creative moments was a universal ignorance of those basic questions about what human beings were from the standpoint of our evolutionary heritage. How could they not understand that fundamentally we were biological creatures? There was little secular curiosity as to where we had come from: What were our possible destinies? Was there widespread and serious scientific study as to *why* there were vast differences in human literacy, intellect, behavior?

These differences in complexity were far beyond the simplistic ideological rants of 'racism', 'sexism' and the deprivations of the so-called 'people of color.' In retrospect, so similar to what went on in the early medieval universities or under the totalitarian governments of the twentieth century, so-called intellectual freedom in the higher institutions of the 'liberal' West had placed such research and discussion essentially under interdict. They were philosophically compromised.

By the end of the twentieth century it should have become clear from the accumulating power and intellect of the Northeast Asiatic ethnicities, and the many millions of creative and powerfully functioning humans of the other ethnicities of color, that color and race were essentially irrelevant to the larger meaning of humanity's heritage and destiny. History was moving. As had occurred so many times in the past, the people, drugged by media propaganda, played with their electronic toys, the leadership—politicians, financial speculators, and entertainers—gathered up the wealth.

Human history unfolds its own story, rarely intimidated by current 'group think.' Civilization was crumbling within its own historical logic. It did not happen surreptitiously. But the masses were instructed not to worry. All that was happening was supposedly a mere momentary blip—'This system of life that we have prepared for you is still the best of all possible worlds.' And of course this dynamic did not meet the Darwinian requirements for positive natural selection.

· 6 ·

BRAIN POWER

Genius Juice

In this chapter I will briefly relate a story that was spread about at the early meetings in Geneva. First however, I want to discuss the context in which this anecdote occurred.

These were conclaves concerned with the path to take into our future. The search was for scientific and intellectual consensus. But there was also a sensitivity to the fact that these research groups of ever-changing contributors, all unified in consensus, had to be reinforced by powerful figures in the various communities. They, in turn, would have to cede power to a new international overseership. The story that was being passed around became the backdrop of the debate over priorities. What should we place first on our agenda in attempting to right the sunken ship? Most feel these meetings took place about 2170–2180.

The choices as to where to begin to intellectualize the efforts for renewing our civilization centered around issues of democratic living, levels of cultural advance and debasement, scientific productivity, population reduction, intellectual capital. As I have mentioned earlier, there was already much freelancing during this period in terms of genetic material being exchanged, purchased, distributed, and shared among our species, given the differing ethnicities and racial components. This widespread trade in sperm donations was universally viewed as increasing brain power.

As I mentioned earlier, this process amounted to either sexual relations or artificial insemination. Again, the inducement was at first monetary. As soon as educational levels in the children produced seemed to soar, demand by the various communities began to outstrip supply. There was another additional but importantly persuasive argument to the citizens of the various ethnicities who might benefit from these new genetic elements. As the national leadership groups argued in many community venues, the genetic contribution from especially gifted males would add great potential talent into the indigenous societies but without the possible cultural traumas that large-scale immigration of foreigners could produce.

In a word, the issue of acculturating new genes attached to their old cultural personalities would now become moot. The children born of these experiments would immediately become full members of the community that they were born into. The phenotypic racial or ethnic hybridization that would result was irrelevant to the leadership groups. In short, the issue of racism also had become moot.

As humanity emerged from the fog of war, disease, anarchic chaos and violence the reconstruction of a world community seemed to need this unified high-level brain power to re-master the old skills and scientific technologies. This had resulted in a relatively small proportion of truly modern humans during the early twenty-first century. To be left behind in this twenty-second century process of modernization as it developed was now seen as the pathway to serfdom and any new forms of enslavement that might stimulate the perverse consciousness of some of our brethren.

The story that I have been leading up to involved a warlord in early to mid-twenty-second century Central America. The humorists in Geneva probably made him into a caricature. He determined that he needed his own private supply of sperm from elite men. On the advice of his shrewd and self-serving followers it was decided to invest the pittance of gold in their vaults and have specially chartered aircraft fly down from North America a refrigerated supply for distribution to all the ladies of his even-then considerable fiefdom. The children with higher brain power were to create the intellectual/economic/military conditions for domination over all of Central America and perhaps over even lands further south.

A probably apocryphal coda to this story is that the warlord demanded for his gold that the desired substance be purchased only from interns of the Harvard Medical School from all racial and ethnic groups.

Apparently no one in Geneva could verify the results of this capitalist investment. By the time this progeny cohort could have matured and have benefited from the educational establishments he had presumably planned for, the international Secretariat had already absorbed the significance of these clues. Sperm donations were working around the world. There soon were many other ad hoc examples of this type of response. The various communities of the world had gone ahead of Geneva in attempting to realize the hypothesis concerning the relationship between high human intelligence and those economic/scientific/technological benefits to be expected as a result.

In the end our World Society predecessors were in agreement. In the long run we could not expect to enact demographic discipline until we could balance death control and birth control. They had evidence that communities of high intelligence and advanced socio-economic social conditions can be alerted to the impact of family birth numbers on future economic and social consequences. Also, as noted above, scientific advances are definitely correlated with high intelligence and advanced educational achievement in any society regardless of racial or ethnic heritage.

Democracy, the long evanescent dream of political thinkers and utopians, desperately needs a smart, involved citizenry. As Plato noted thousands of years ago, democracy cannot last. It leads to oligarchy, autocracy, eventually tyranny. Why? Humans often think "with their blood" rather than with their cortex. Populist demagogues always sing sweet songs and have charismatic personalities. They always say that they want to help the people, but they still enriched their friends, accumulated totalitarian powers and clouded the airwaves with the rhythmic beat of religious or ideological oratorios.

The following proposal began to gain overwhelming assent: Unify our species. Make us all 'Cro-Magnons' regardless of the color of our skin, the speech emanating from our tongues, the geographies and ecologies contributing to our life expectancies.

A Unified Species

Our founders had then to think about the 'how' of our decisions. In the early twenty-second century there still existed great differences within humankind, educationally, religiously, culturally. Clearly, if the concept of raising all human intelligence up to its highest sustainable level was going to work, it had to start experimentally and be worked out rationally and humanely over a long string of generations. Truly, the mysteries of our genetic heritage had not yet been unraveled.

Haphazard research during previous decades, attempted through a variety of gene therapies which altered our genetic profiles through biological and medical manipulation, had led to many unintended and disheartening consequences. The traditional interpersonal sexual transmission of family and community genetic profiles still seemed to be the safest and most reliable approach. However, artificial insemination and prenatal genetic analysis, all part of genetic innovations achieved well before the cataclysm, held out the fastest way to achieve a worldwide socio-biological program.

It should be stated that in this period, early to mid twenty-second century, when the reclaiming of modernity and scientific/medical advancement required highly educated humans, these ad-hoc experiments by tribal leaders, warlords, even modestly self-governing communities still took place. There still was ongoing tragedy. Few central governments were available to dole out welfare and distribute survival to one and all. We were living the Darwinian scenario in which the able struggled to compete with the less able, and rampant starvation and the withering of communities, infanticide, and the death of unwanted children through neglect were the norm.

Fifty years later, consultations with the leadership elite of various ethnic groups in different continents revealed to scholars in Geneva that the horrors of the preceding centuries had cleared the air of earlier ideological intransigence. These forms of irrationality existing in the traditional political and religious blocs had once stopped science in its tracks. The new generations now seemed to be hungry for factual routes towards secular salvation.

Male and Female

It is appropriate to discuss the thinking that underlay this proposed international undertaking. Human phenotypic physical and behavioral characteristics derive equally from the genetic contributions of male and female parents. Without the sharply restrictive social policies that attempt to re-determine male and female social roles, a new beginning was possible. In this time of extreme scarcity of resources it was now possible to speak of our evolutionary heritage of human sexual dimorphism.

Our view was to ensure that both men and women would contribute complementary social roles to their respective cultures rather than attempt to mimic each other. No longer would we attempt to manipulate men into being women and women into becoming men. As Rousseau had written many centuries ago, this would and did lead to the degeneration of both sexes. In the early twenty-first century version of modernity what we were mostly seeing was the transformation of the female profile into the male—in sports, vocations, and education. Of course this had an impact on male behavior too; note the proliferation of male homosexuality in the Western nations.

In so far as intellectual functions are concerned there were traditionally important differences. Over the course of millions of years of human evolution and changing brain size and function, the male had consistently had to take on the responsibility for hunting, fighting, aggressively searching out for more hospitable ecologies, ever on the move. The female, on the other hand, was shaped by the need to care for neonates that were completely dependent on the mother at birth and for increasingly long periods afterward.

The human female so radically reshaped from the traditional mammal patterns of sexual periodicity developed, along with several anthropoid species, a constant availability for sexual relations and the inevitable pregnancies within the year. The nine months of gestation which the modern female experienced was not a time when she was able to maintain activities that paralleled those of the male.

Nature demanded of the human female a psychological and biological accommodation to the need for species survival. This meant that she had to have a persona of protectiveness, nurturing, emotional stability. In addition it was required that she be strong enough to allow for successive births of a delicate and completely dependent infant. Anthropologists had long observed in the fossil remains of the more advanced Cro-Magnons a large proportion of very young deceased females often along with their new borns.

This bio-social profile of the genders could be extended into the intellectual domain. Thus we see in a variety of testing formats yesterday and today that the male profile varies far more greatly from the norm than does the female. The standard of deviation of the male is c. 15 points, the female c. 12. This means that males contribute a far greater percentage of mental and psychological disabilities than females. On the other hand at the upper ranges of intellectual and educational achievement males predominate.

To elaborate simply, in societies of a lowering intellectual profile it will be the males who will hurtle the society downward, the females remaining closer to the norm of their group. In societies where the intellectual profile is heightening, you will find more males at the genius level than females. Conversely, as we look at mental retardation and institutions for the criminally insane and other disabilities of the mind and behavior it is always the male who predominates, defining our picture of mental pathology as compared with females.

This is to say that in a modern society where physical labor is practically non-existent for both males and females you will find the female mind at very great advantage in a variety of professions, even leadership positions. But to be sure it is the highly intelligent mother's son who will become a whiz kid in the sciences, the abstract arts, and other areas of critical national and international survival and progress. These facts underlined our national and international need for highly cognitive and creative humans.

It was the savants in those early days of the Geneva Directorate who decided to distribute the sperm of the most highly educated and creative males on the planet to carefully selected females throughout the world. How was this accomplished?

· 7 ·

ROAD TO SOCIAL EQUALITY

An Intelligent Citizenry

Another famous story told and retold was about the political ship of state, a series of meetings of the scientific and legal committees of the Directorate in Geneva, c. 2180. Scientists described the manner that genetic material would be distributed, to whom and where. As the particular speaker was exhibiting demonstration vials that were being stored and then distributed to states and nationalities that had agreed to participate, the voice of an old judge from Kenya of European and African heritage was heard. I paraphrase:

> You are showing us different sample vials of genetic material from various national and ethnic sources that you intend to infuse into very different ethnic and national locales as an experiment, I presume. Why not utilize larger vials and mix up the contents (sperm) of all these particular vials?

They say that this meeting then broke up in a roar of laughter. Actually it was decided to go halfway down the road that the judge suggested. They would indeed blend the genes of many talented men from all over the world in some of their trials as a control. Other ethnic and racially homogeneous genes would be directed at particular different but somewhat homogeneous communities and the outcomes checked.

The women in the communities who had volunteered to take part in this program conformed to the programmatics suggested by several Northeast Asiatic historians. The ancient Chinese road to high intellectual achievement was characterized by the privileges enjoyed by the highly tested and elite mandarin bureaucracy of their imperial system of governance. It is thought that the polygamous marriages of these eminent intellectuals and the many children they produced were the key to the high intellectual civilization of the Han and their subsequent dominance in the twenty-first century before the 'crash.'

As hypothesized by these Asiatic scholars, the winners of the competitions for intellectual excellence usually sought wives who were the progeny of up-and-coming mercantile and agricultural families. Although the daughters of the elite aristocratic classes would rarely have submitted to bluntly polygamous marriages, the ambitious middle-class achievers were anxious for their daughters to bear the children of men with these unique minds and their powerful political and economic status and connections.

The women in the various communities of the twenty-second and twenty-third centuries who agreed to be participants in our great experiment were themselves tested in regard to specific talents as well as carefully scrutinized as to their personality and various other psychological traits. Naturally they and their families were paid from a special fund that was now available to the World Society from a now rapidly expanding and economically stable international membership.

The women lived in special communities overseen by elderly women with special medical training. Some of the communities allowed for voluntarily sterilized husbands or male companions. There was much local control over the social life of the chosen females. A few of the differing national political leadership elites went back to Israel kibbutz patterns of inter-community independence and economic integration. Needless to say, as mentioned earlier, two generations on, the demand for the sperm of the elite almost overtook the supply.

Socio-Biological Road to Equality

The socio-political vision at the basis of this program of intellectual equality for our world community helped to persuade many different ethnicities that they should give it a try. The motto seemed to be: if any national political community is unhappy about the state of their educational, economic and social condition, let them not accuse the world around them. Try to do something to better your condition. Give biology a try!

And, of course, after even a modest two generations, approximately 50 to 60 years, it was clear a new generation of youngsters had come forth into their respective communities who were ever more able academically. They then became responsible contributors to the societal affairs of the respective communities of which they were now members. Naturally there were many dimensions of the program which needed on-the-spot corrections. Some of the women had difficult pregnancies; abortions were needed. Others changed their minds after deciding that the payments they received were not equal to the isolation or sense of uniqueness that they were now made to feel. Always, new recruits were found; the large waiting lists grew longer over time.

As I write, these unofficial and official programs have been in effect almost 150 years. Today there is so much intellectual power throughout our world community that each ethnic and/or national unit now sources its own program, its own internally and/or externally produced vials of cultural power to distribute if it so wishes. The awareness is clear. Not all intelligent, well-educated and talented people muster up to the psychological requirements of modern responsibility in the practical dimensions of life. And there is much unique talent at somewhat lower levels of educational and intellectual achievement. The most innovative philanthropists of my own day are not necessarily the most brilliant intellects of our time.

Additionally, what we have learned in initiating a revolutionary genetic amalgamation of our people is that intelligence is key to what the nineteenth-century socialists envisioned as the doorway to a classless world society. The genetics of human talents are now being so emulsified into the respective communities of our planet that, when the baton of leadership is passed from one generation to the next, we see the dividing line between great and modest social success as wavering and vacillating amongst families and communities from generation to generation. It has made for an interesting parade of elites over these years. One hopes that prefigured dynastic succession will not exist in the future.

End of the Races

In the beginning, the North Europeans were very hesitant, if not recalcitrant, about agreeing to share in the genetics of other races and ethnicities. Nor were they enthusiastic about donating the genetics of their finest to the rest of the world and to future competition. They had ridden out the century and a half of horror fairly well but still, they were scared.

The remnants of the indigenous African Negroids were also quite concerned about seeing their one-time vision of greatness as a people of wealth and progress diluted by this probable infusion of genes. The emptiness of Africa and the richness of its lands seemed to augur a new beginning for this continent. But there were, as I have noted, early dissenters from this ethnic sense of exclusivity, precisely because of the already pluralistic racial heritage of its people. In a quick transition Africa did become a multi-racial, multi-ethnic continent of great success and probable future wealth.

The Africans were persuaded by their own resident intellectuals, many of them already racial hybrids. I can say personally from the distant standpoint of over one hundred years that a hybrid African great-grandfather of mine argued in the affirmative for this program. One of the more persuasive arguments was that an ancient African hero, Nelson Mandela, was himself a hybrid of two indigenous African races, the Capoids/Boskopids of the click languages and the more numerous Negroid Bantu.

But in truth there was already much migratory Caucasoid hybridization into these two indigenous African races. On the continent of Africa there were in the north well hybridized Caucasoids with migrating Negroids from the south and indigenous, if remnant, Capoids. Here we include a large swath of geography from what was once Egypt to Morocco and then south into lands once called Ethiopia, Rwanda and, of course, South Africa. Even before the cataclysm, South Africa was well mixed genetically because of its intrusive Caucasoid populations of Dutch and English overlords, as well as the Bantu peoples, who themselves were of West African origins.

For the Northern European Caucasoids, the change in their attitude came with the decision of the Chinese nationalities, also the Koreans and Japanese, to contribute to this potentially massive and revolutionary restructuring of the human species. 'We are already tri-racial hybrids,' was the chuckle from Beijing, Seoul and Tokyo. 'We won't mind if we see new slant-eyed smart ones in Africa or in the Americas.'

It took only two generations for the products of these experiments in long-distance miscegenation to realize the potential success of the program. The human race could receive a new integration of human types characterized by an increasingly higher level of practical as well as theoretical intelligence. The program did not produce pathological freaks.

Rather, what the Secretariat was achieving with the assent of an increasing majority of states and nationalities, was a speeding up of an evolutionary process that had been taking place in the human species for several hundred thousand years. (I will amplify on this in a further chapter of this memoir). Nothing new was occurring except for the deliberate infusion of "highly cerebral" "g" factored genes from all the races and ethnicities and then into all the races and ethnicities. The miscegenation of the race would have happened in all cases, and probably with many more horrors to come. But now, the facts on the ground, so to speak, had created a truly democratic, and relatively uncoerced series of decisions by the existing citizenry of every national unit, in the name of a new beginning for humankind.

Some people asked, why the racial focus?, why the rush? The answer from the very beginning was sure and clear. The most important dimension of human nature, its many ongoing mysteries lies below the skin. Our phenotype represents a memory of what we were long ago on our way towards sapiency. Race carries with it ancient and now irrelevant remembrances of the past. By integrating the genetics of the various residual races and ethnicities we are not turning our backs on our nature's evolutionary history and destiny.

Certainly the dark skin and distinctive features of the Australid and Negroid races is a carryover from their long adaptation to the tropical environments in which they lived when discovered by other races as far back as the sixteenth century.

And it is true that the so-called Mongoloid and Capoid races have been exposed to a somewhat intermediate climate and geography. Some scientists speak of the Capoids, sadly, as a now vanished race. However, very much like the Australids, both racial traditions have their hybrid descendants well represented in our new universe of genetic amalgamation.

The Caucasoid race has long been a resident of the lands of the pale Northern Lights, not recent migrants from Africa. And they are well represented by many primitive exemplars, the Neanderthals. If the Cro-Magnons with highly cerebralized brains were Caucasoids, it only testifies to the drama of survival that long ago took place amidst the frigid glacial advances in these latitudes.

So, much political antipathy in the recent centuries was caused by the seemingly sharp social and economic differences supposedly epitomized in race. The hybridization of humankind had already speeded up in the two centuries before the lights went out towards the end of the twenty-first century. Why then, should we not eliminate race from the repertoire of human political and cultural discourse? It only represents surface structure.

Beyond Race, Culture

What we humans need to achieve over the next millennium or so is an ever-deeper understanding than we now have of the human brain and our nervous system. Racial difference are ephemeral. We are now, as I write at the end of the twenty-third century, enjoying, even playing with the interesting cultural and intellectual nuances that this out-breeding of humankind has brought about. Not insignificant is the sense of beauty that has expanded our repertoire of artistic renderings of the human body, the human face, hair and skin color.

What is most important in this general intellectual advance is the fact that no group of people having certain phenotypic characteristics can be labeled as a separate race. Neither ethnicity nor nationality can now be defined as 'such and such' by their external physical attributes, or, of course, as being better or worse than others.

From now on we should speak of differing ethnicities, involving different cultural/symbolic shapings of our conjoint human experience. Today we're already beginning to treasure these differing cultural heritages, here, since the beginning of the 'great mixing.' A way of thinking about our international situation is to contrast the 'different' phenotypes of historical England and France. Not much there.

Think about the diverse cultural, scientific, esthetic contributions made by each nationality. There is much there to be treasured by all of humanity. Examine the cultural contributions of England and France in their music, art, even scientific approaches to the study of nature. Extend such comparisons to those differences between the southern state and the people of Shandong City and those of the north in Harbin, both Chinese in racial and cultural heritage.

Humans are not going to look the same. Each community will be likely to inbreed and shape themselves into a similarity of appearance, yet cross-breeding will always exist between nationalities.

· 8 ·

THE LONG ROAD TRAVELED

Reason Shipwrecked

The secular mind liberated itself from myth and the gods in Greece, approximately 2,500 years ago. Socrates, Plato, and Aristotle committed themselves to an understanding of the world on the basis of the factual knowledge existing in their time. Socrates argued from the Agora, the central meeting place of ordinary folk. Not until his rump judicial trial in Athens and execution did he become a public political figure. His service to humanity was to question and attempt to unleash reason from its psychological and emotional fetters.

Plato schooled his fellow Athenians and others throughout the Hellenic world who would listen and learn to the larger intellectual significance of the Socratic vision. Several attempts were made to put the new secular understanding of men and society into a visionary political context. Plato did not want to see another tragedy such as the Peloponnesian Wars that essentially destroyed the dominance of his own polis, Athens, and brought an end to the Periclean hope for a great Hellenic civilization. Plato, a utopian, in his greatest work, *Republic*, consciously destroyed his own utopian vision. Even his golden-souled rulers erroneously miscalculated the time frame of nature's order and thus destroyed the ideal Republic.

Aristotle, a non-Athenian from the 'countryside,' was not inclined to the utopian vision for humanity. His route was through science and logic. All that we could say about human nation-building and the hope for peace, prosperity, stability, he lectured, could never be absolutely framed in policy. The human universe as contrasted with the material universe of matter and life could be understood in probabilistic terms only. This was true even in the limited monarchy which he hesitantly put forth to his students as the best of all possible solutions.

In the following millennia we have had other attempts to place humanity into history and reality. Many religious thinkers into the so-called Middle Ages such as Thomas Aquinas worked over the classical Greek vision, the intangible god-head, Aristotle's 'unmoved mover,' the logical *Theos*, becoming the summa of Christianity's trinity. Also persuaded was the Jewish Moses Maimonides and the various Islamic thinkers of this era. The intellectual dream was to create a mirror image down here on Earth of the heavenly beauties of the immortal spheres.

In the modern period, figures such as Baruch Spinoza, Thomas Hobbes, Adam Smith, and Karl Marx extrapolated from the scientific secular world those movements of stars and earthly bodies. The goal was human moral, social and economic redemption. Implicit was a utopia that had its roots in the firmament of knowledge and power that science had released to the human mind.

As we have seen science set off social and psychological forces in the human soul that ran away from reality and crashed the secular vision. At the end of its 500-year extravagance of attempted control over nature, it lost control. The irrationalists were in part to blame, human reason was weakened by the masses of flesh, ignorant, emotionally susceptible. What mentation resulted could hardly call itself a product of *Homo sapiens*, no less *Homo sapiens sapiens*.

Just consider all the purveyors of snake oil in those vulnerable centuries. The knowledge about man and nature was there. Karl Marx was a believer in the factuality of such analyses. In fact, he was a follower of Charles Darwin. By the twentieth century his followers had created a great mythos built on his understandably erroneous theoretical economic presumptions. These conclusions led those who should have known better into an ideological crusade of violence. The fascists also did violence to their own deluded.

And in the background were the billions intellectually or through lack of education incapable of understanding the workings of nature. So as with the Pied Piper of Hamlin they followed these perennial purveyors of fairy tales, doling out their supernatural balm from their cushy churches, synagogues and mosques. But the worst of this behavior took place into the modern era, in the twentieth and twenty-first centuries. How could the world not fall apart in this clash of the smart few with the ignorant masses? To say the least humanity was linked into a great conceptual disconnect which could only lead to the tragedies the species has experienced in the past several hundred years.

Mind you, reader of this missive, I am not dismissive of the religious factor in the human symbolic structure of thought. In sections that I intend to present here, I will show how the modern perception of human nature does give an important place to those inchoate mental demands that we describe as the religious. But they are at the present time severed from secular socio-political policy making.

So here we are several hundred years into an attempt to right the ship of reason and build an enduring context for human life. There will be disciplined change in the symbolics of meaning. And we will call it 'progress.' I hope it will not be accompanied by the bestial horrors of war, genocide, persecution, urban degeneracy and poverty—so much unnecessary human suffering.

Our Secular Perspective

Much of the perspective which I am going to relate here was part of the scientific consensus before our twenty-first century civilizational collapse. The problem was that it was not made available to an educated public so that this knowledge could become part of a basis for national and international policy making. The scientific community, especially in the areas of human evolution, psychology and genetics lived in fear of the ideology that dominated the governmentally authorized 'truths.' Factual knowledge succumbed to the 'mainstream media,' the educational system from kindergarten to higher university research. Orthodoxy prevailed, call it religious creationism or the other side of the political spectrum, sociological creationism. Truth was ground out between these enforced mythologies. After all, science then depended on the governmental spigot to survive.

There is nothing like disaster to clear the air of assumed truths, as the Germans, Japanese and Russians learned in the second half of the twentieth century. Today there exists a dialectic in the human social sciences stimulated by the dissemination of evolutionary thinking. We continue our search for the meaning of our essentially biological nature. What we call culture, society, is a product of our biological, behavioral expression. Believe it or not, this is not a revolutionary concept. Darwin and others understood these issues in the nineteenth century.

The forces of nature produced the biological world, which we humans have abused and destroyed, and have over the past several billion years produced this creature we call *Homo sapiens sapiens*. This understanding of humanity's place in nature and on this planet should have been front and center in the education of the most modern people from the nineteenth century on. Instead we have suffered layer upon layer of religious and ideological obfuscation and therefore opened ourselves up to the disasters we have foisted upon ourselves.

It is a long, fascinating, complex story. Here I will relate a very short version of the facts that are part of our World Society governing philosophy.

Animal Determined

Life on planet Earth has taken many pathways to avoid the energy degradation predicted by the second law of thermodynamics. Our pathway as an animal, phylum chordate/vertebrate, has epitomized movement, predation, sensory awareness and reactivity to the environmental-ecological changes that nature has thrown at life. The time span here constitutes 1 to 2 billion years of animal/vertebrate life and evolution.

In fact one can with a bit of hesitancy say that the conditions of physical and biological changes on this planet throughout its more than 4 billion years of existence argues for a determined process of extrusion of a form of animal life that was to become human. This creature was to epitomize this determinacy in the vertebrate taxon, high cognition, thinking, adapting within the generations. Why? To survive the changes that nature threw at life forms practically every year.

There is, therefore, an evolutionary inevitability here on planet Earth for the coming into adaptive and selective success of a highly intelligent survivor of the environmental challenges to life. This creature, *Homo*, should be capable of altering his behavior within the generations without the need to profit by a chance genetic mutation, here giving the species the chance for another positive accidental role of the dice. Nevertheless, we have no evolutionary assurance that *Homo sapiens* will prevail beyond an evolutionary specific period of time.

Yet, in retrospect, the adaptive and selective characteristics of a biological creature with a large brain and an ability to think through the challenges of brute survival while bringing its young to reproductive maturity seem to have been a persistently and positively selective attribute of those antecedent creatures, vertebrates, mammals, anthropoids from which we have descended over these many millions of years.

Coming of Man

Even in the 'old days' the issues of vertebrate and then hominid evolution were ongoing science and not subject to major political interference and suppression. When we arrive at the question of the final stages of human evolution the silent binding of what was called political correctness dominated the late twentieth and early twenty-first centuries. Today, the old controversies have been long put to rest. The East European, Chinese and Viet Nam revolutions against raw communism as well as the ethnically secure Koreans and Japanese helped to push away the fear of retribution and economic destitution that many in the Europoid scientific community had worked against.

Simply said, this relatively recent revolution in the creation of *Homo sapiens* occurred in Euro-Asia and not in Africa. The servile ones were wont to serve up gratuitous scientific hokum to the mass media and their overseers in government. Predecessors of *Homo sapiens sapiens*, humans such as the Neanderthals in Euro-Asia, were very heavy of bone signifying, from the traditional evolutionary point of view, a culminating point of success in the evolution of a line of animals.

For example, the Miocene apes of Africa, c. 20–12mya were mostly gracile types, more like gibbons and monkeys than big guys like the extant gorilla. The earliest hominids, Australopithecines and the early forms of *Homo*, were gracile, except for one group of robust Australopithecines. The controversial Flores hominid fossils found in Indonesia in the early twenty-first century were estimated to have survived until to c. 18–15k BP, and in structure were very primitive, with a very small brain even for its tiny gracile structure. It is an exemplar of a very early migrant out of Africa.

Homo rudolfensis, "1470 man," in East Africa is a possible precursor to Cro-Magnon at c. 3–2mya. It is a large-brained relative of *Homo habilis*, the earliest true *Homo*. After these two and other human precursors the next stage of the human enlargement in brain and bone is summarized in creatures labeled *Homo erectus*. The evolutionary sequence of these humans is still subject to conjecture as much time has flowed over the horizon between these fellows and the final extrusion of Cro-Magnon, *Hss*. The ecological context of the origin of *Hss* has to lie in the pulsations of the Pleistocene Ice Ages, from 1.9mya. This era, almost two million years of great climate change, probably had an enormous impact on the survival of various lines of animals. What happened during this evolutionary act of creation was an event of the north, not Africa.

Cro-Magnon

The thinking of a psychologist colleague highlights the spurious controversiality of this evolutionary conundrum. He is of greater indigenous African heritage than I am. He also claims West Africa as his origin. I argued that he looked more East African, Somali, Eritrean (significant Caucasoid genetics). His comment on this old controversy of Cro-Magnon origins: 'Poof, it doesn't matter, those guys were just as subject to regression towards the mean as we Africans.' He was pointing to some of our other colleagues, more European or South Asian, whose children had not lived up to parental expectations.

Of course I am proud of my allotted two and my psychologist friend's two. Our progeny and their children hardly show any marks of such 'regression.' These four, all successful contributors to their adopted nationalities, have married into these new ethnic horizons!

Many hundreds of thousands of years ago the Cro-Magnon, *Homo sapiens sapiens*, unexpectedly arrived. But they only showed up as a developed cultural and technological reality from c. 50, 000BP (Before the Present). They were extraordinarily different from the existing 'form of humans.' Their skulls were not much larger in size than the Neanderthals, but the structure was different. The frontal brows vaulted upwards. The bones of skull and body were delicate by comparison. This human was taller and seemed to have walked as we do; the various heavy bone structures of face and skull no longer there, only an upward sweep, then a great protuberance soaring above the delicate facial features. They were the modern Europeans, Caucasoids like the Neanderthals, but with a wholly new makeover.

Evolutionists speak of a genetic bottleneck, ostensibly several hundred thousand years ago. One could also call it a funnel out of which a small number of variant genes came through a filter of restraint and then seemed to have surged outward, creating a large and variable population but still of genetic consanguinity. It hints to us that the revolutionary human brain was the evolutionary result of unknown selective stresses which eliminated many other human genotypes and allowed this unique profile to exit the "bottleneck/funnel" and become the dominant human genotype in our world.

What is significant in understanding the new behaviors that seem to explode out of this human mind is that these proclivities of mind are in no way related to the traditional sign/signal behaviors of instinct that had guided animal life for well over one billion years. Scientists call this progression of brain growth and then the explosion in size and power of the cerebral cortex, 'orthoselection'—natural selection in a straight line.

Those new genetically controlled behaviors, thinking through situations slowly and thoroughly to make a behavioral choice, were often successfully expressed down the line in time (time binding). The key to successful adaptation and natural selection was to ensure the successful nurturing of the young until maturity. Over a possible time line of hundreds of thousands of years and during a period of northern Ice Age climate and ecological instability, a potential outlander of behaviors, childlike and aged/wisdom, complementary youthful physical attributes, were all catapulted into selective reality.

This power we today call abstract thought, decision making in time and space, away from the originating sensory inputs. These inputs needed evaluation rather than immediate response.

· 9 ·

HOMO SAPIENS SAPIENS: EVOLUTION'S MYSTERY

Disturbing Lecture

The World Society moved to the Nairobi enclave in the former nation of Kenya in 2200. By then there was a plurality of political units around the world functioning with enough capital/economic strength to fund this move, its physical development and sustenance. I joined the Directorate as a glorified waterboy in 2230 at age twenty-six.

Shortly thereafter I attended a disturbing lecture, part of a series of lecture presentations and discussions. It was given by a prominent scientist in his/their study of human evolution and human nature. Although this lecture paper resides in our archives, I use my notes to paraphrase. I believe this lecture to have focused on the essence of our task to guide our world community into the mysteries of human nature. Inevitably this knowledge has to lead to the question of the policies we need to introduce to maintain our balance on that winding path that this mysterious biological entity will take into an unknown evolutionary future.

The lecture concerned our basic knowledge of the expression of this human nature as hinted at by our understanding of the thinking and behavior of the originators of cognitive human civilization. These originators, albeit of a nonliterate civilization, were the Cro-Magnon peoples of Europe.

Abstract Symbolism

I paraphrase the lecturer's remarks:

> We as with all higher animals/ mammals receive our survival information through our senses. The senses act radar-like, quickly pouring out neurological queries from our cortex to translate the raw sensory information. The human brain thus transforms this raw data into symbolic questions, first to find meaning and then to create ever-more abstract interpretations of meaning. These questions are not to merely identify sounds, visual objects, smells, or touch or taste.
>
> Rather, the questions the brain exudes are far more penetrating and significant. Is there a causal imputation here? Can we establish in law, a regularity principle which will connect up with other cores of organized sensory information? Of course the point of this searching for a 'law' is human survival. What kind of law for survival? The cortex always asks of the avalanche of sensory data, a symbolic, conceptual question.
>
> These two questions epitomize the great human deviation from our animal origins. Substituting for the instinctual regularities that animals expect from their own signaling neurological survival system, the brain is now emptied of specific behavioral triggers from the incoming sensory flow. Instead it throws out an almost endless series of causality questions that we direct our sensory intake and interpreters to answer.
>
> Our brain's cortex acts so, automatically. Instead of automaticity of behavioral responses, fleeing, attacking, hiding, our brain asks, 'What goes'? What are the laws, the principles and regularities that we must understand and take account of, so that we can live and adapt to this always changing environment? Put another and simpler way, 'how do things work'?

A Practical/Impractical Brain

The lecturer was, of course, referring to the brain of modern humans, the Cro-Magnons in origin and of us, the hybridized *Hss* today. The Cro-Magnons appear to be the first bearers of this new revolutionary brain. Their behavior indicated by the cultural relics they left behind reflect its mysterious workings. When they arrived in Europe some 45,000 years ago they displayed for us, from the Balkans, Russia, and finally into France and Spain a fully articulated and unified culture called the Aurignacian.

Their movement westward into Europe, probably during a mild interstadial, to where the hunting was good, already displayed a highly developed technology and art. They probably had long since mastered the techniques of hunting and the gathering whatever autumnal fruit and vegetables would grow in this dynamically changing Ice Age ecology.

The nature of their culture implies that they already were beginning to put 2 + 2 together when it came to figuring out how non-instinctual survival on this Earth worked, how a community must think about these external causal relationships. For example, survival on the hunt did not require the close-up killing of a large mammal. Just by wounding a large bear, a mastodon, horse, or reindeer the hunters could follow the bloody trail on the snow or earth, eventually surround a dying or dead animal, quickly skin it and bring back the hunt results to the encampment.

Their efficiency in learning the rules of the hunt and the climatic and migratory regularities of their ecology allowed them to live well and grow in numbers and security. On the other hand the mere several thousand years of this pulsating Cro-Magnon expansion pushed the frustrated mentally lethargic and finally desolated Neanderthals out of successive habitats and relatively quickly into extinction, c. 40,000 BP. This evolutionary succession did not take place without the addition of some Neanderthal genes to the Cro-Magnon clade and eventually to the modern European. The same process of hybridization has occurred with the Tasmanian natives who became extinct in the late decades of the nineteenth century. Their progeny has fructified the incoming and now-dominant demography.

The world of economics relationships, the environment from which we attempt to make a living, is relatively secular. It changes, but in pulsations that the modern brain can and did master—to feed the tummy, to clothe the body, to sustain fire, to warm the cave. Another side of the brain seeks answers to two basic questions that determine human survival. This is the emotional component of our mammalian heritage, our limbic system. The lower brain controls and disciplines basic autonomic animal functions, but its anthropoid components pour vast emotional energies upward into the cortex for interpretation and expression.

One sees elements of these passions in the tools of the Cro-Magnons. These tools are in functional variety vastly beyond anything other humans previously created. They are not merely practical but are so fabricated as to clearly demonstrate the conscious awareness by this brain of beauty and the need to express the beautiful in the lives of these people.

The esthetic element shows through in the fact that some of these tools made of bone, ivory, stone, shale were too delicate for practical use. The passionate mammalian energies that the brain has inherited and amplified from the mammalian core are here powerfully endowed with a cortical dimension. What we call beauty reveals an imaginative element in the human psyche. What 'Darwinian' practicality is embedded here? It could not have occurred if the Cro-Magnons were obligated as are most animals, to spend twenty-four hours a day to find food for minimal survival.

Along with the beauty of their tools, as with all moderns, their decorative jewelry for the male or female reveals highly developed skills always undergirded by the search for beauty and the energetic rewards of craft fulfillment. Having gained the time and space from the practicalities and mundane success of the hunt, the burning symbolic energies of these mentally powerful people were transferred into a seemingly never-ending passionate search for the art object. Once the challenge to the questions of everywhere and always with regard to the laws of the hunt and economic survival were absorbed into routine knowledge expeditions, the always creatively challenging art object became the new focus of these passionate questions of meaning.

Think of the Venuses that were fabricated throughout Europe. Of course these were the products of humans who came through that mysterious selective funnel many hundred thousand years before they created this nonliterate civilization. The Venuses were carved from a variety of materials, sometimes of fire-hardened clay as in the later pottery of nascent urban communities. They were probably used as mostly hand-sized tokens of memory, objectifications in art of their throbbing sexuality, perhaps the love of the mate left behind as the males wandered forth on the hunt.

The Venus of Laussel (France, c. 27,000 BP), the relief carving of a nude, possibly pregnant, female carrying and looking towards an animal horn with thirteen decisive vertical incisions, has evoked much interest and speculation. Is this carved "Venus" related to a temporal evocation of the rhythms of insemination, pregnancy and birth, or is it an astronomical abstraction of the phases of the moon, thirteen days from first crescent to full moon? Or does it attempt to relate the crescent of the animal horn she carries to the times during the sidereal year that the crescent will appear in the sky?

This is one of the many chronometric evocations that the Cro-Magnons left on numerous plaques, which are seemingly attempts to symbolize the regularities of life, from the female menses to the changes of the seasons, perhaps like the marks on a hunter's rifle that indicate kills. Long before the autumnal Ice Age climate changed to very harsh winters and soft summers as major portions of these populations made their way south, a human adaptive revolution had been fully realized. The highly encephalized Cro-Magnons of Euro-Asia were probing the abstract symbolic potentialities of the brain, registering its spontaneous emanations, always attempting to mentally order the pulsating concreteness of the sensory world being revealed to them.

Mystery of Meaning

The Cro-Magnons in their movements west and north into Europe from c. 50,000 BP endured approximately 40,000 years of intermittent glacial descent. We humans recently have experienced almost one hundred years of the dry and cold. It may happen again. But during this most recent period of climate abatement we have endeavored to lift the heavy weight of impotent (if now exploitative) human populations living upon this Earth. There are today many fewer humans on this Earth than in 2070 when the cold/dryness set in. We still have a ways to go in our program. The worry continues because, as we make our plans and then put them into effect for all the peoples of our planet, do we know what we're enacting for our descendants down the road?

The Cro-Magnons over their span of history reveal to us a broad picture of what this new human brain wants to do on Earth. There is no clear biologically adaptive imperative emanating from this cultural picture from long ago. Neither is there a true message for posterity in the urban civilizations, beginning in Mesopotamia and Egypt, later extending itself into India and China.

We still witness ubiquitous passions for war. But there is also the beauty that we see in the art forms of all peoples. There is also the steady search for abstract meaning, the wonder at the events and realities of life that humans have to face on this planet. The search for causality allows us to experience not only outcomes, the consequences of things but also the discovery of the abstract techniques for making our intercession with nature an adaptive one—written language, mathematics, then engineering and architecture. Into this future comes musical notation, the desire to structure the sounds of the voice and the artifacts of musical performance into an abstract intellectual system while yet preserving those deeper rhythmic and lyric expressions of the heart and the blood.

The desire to order the physical world mentally, to order the emotional tugging of the art forms into systems of understanding, beauty itself demands of the mind interpretation. But what does this mental haranguing mean for the survival of the human race, for the biological significance of the workings of this brain?

Behind these serious questions and hypotheses comes that most basic theoretical reality. We are here on Earth along with all the various phyla of life as an exemplification of our resistance to the second law of thermodynamics, 'time's arrow.' Living things pretend to defy this law but only temporarily. The law tells us what will happen in all thermodynamic systems. Our solar system is an exemplification of the gradual giving-off of energies into the empyrean of emptiness, the heavens, the stars of the Milky Way, distant galaxies.

The second law of thermodynamics does not tell us when the final cooling will take place. Nor does it inform us as to what the intermediate events of this process might be, much less the possible time frames of momentary counter events. We, along with all of our fellow biological travelers, are surging in this time warp. For a moment we are free from its determinate predictions.

Then consider the following question for both our short- and long-range future. What does it mean for the future of our species that we have lost those instinctual messages of behavioral restraint that evolutionary experience has imprinted into our genoplasm?

These are the kinds of issues that the planners of the World Society will have to confront. *Homo sapiens sapiens* appeared on Earth as a unique sport. There are no predecessor forms that we can excavate from deep into the earth that will tell us how not to make the mistakes of the past. In this sense we, you, and yours—all of us are on our own.

· 1 0 ·

GOAL OF A UNIFIED HUMANITY

One Government

By the time the World Society bureaucracy was moved to Nairobi in 2200, the political/philosophical battle was over. Those who struggled intellectually in Geneva and even before for a world union, a political unification of *all* the folks, a world government on the planet with teeth, won out.

There were those who waved a flag of potential disaster, even a new coming of totalitarianism, if we went in the direction of political and social uniformity with military enforcement as the backup. It was not 'remember the Maine' as in the old United States' imperialistic thrust but '"remember the League and the UN,' both by the proponents and opponents of a tough love for humankind policy. To be truthful these controversies and debates must have made many an assured head spin around in doubt.

The unifiers countered with the other issues that were already being accepted as necessary policy on this planet. First, we needed to get the population down. By mid-twenty-second century it had fallen from a peak of around 8 billion to somewhere between 5 and 6 billion. Many numbers were passed around as a potential goal. Most felt, considering our natural resource depletion, that it had to decline to around or below what the world population was c. 1900, 1.8 billion people.

Considering the conditions of humankind, c. 2170–2200, many felt that any world organization had to have teeth given the perennial tendency of humans deciding precipitously to go their own way.

The second goal of strong unification was intellectual parity among all the ethnic, national and racial groupings around the world. Considering that this was already occurring out of necessity, no nationalities during these years of organizational formation wanted to be left behind. Indeed, the issue of variable human intelligence had risen beyond the 'Catholic or Liberal index.' The question remained: How can we achieve either of these critical historical goals if there were no powerful centripetal powers endowed to a world government?

The attraction that our historians expressed for a strong world government was that we could have cultural diversity with educational and intellectual equality. The examples of London and Paris were offered up—great productive and intellectual cities but with very different cultural ambiences. Then, the final thrust of the dialectical stiletto. Do we want to maintain the differences we once observed between Singapore and Gaza/West Bank, both with equivalent populations?! The horror of this thought convinced a number of holdouts.

Population sustainability and intellectual and, thus, social parity were the great overriding challenges for humanity. And also, we wanted a world at peace with itself. There was no doubt that to reach the goals of egalitarianism in its most fruitful and meaningful sense we would have to provide the World Society with economic as well as military teeth, But always, in the event of "unintended consequences" we should make sure that these teeth are false, i.e., easily removable. The construction of any world system had to contain within its basic principles, the consent of the governed.

One Humanity

I have delineated in previous chapters the conceptual view of human nature as it existed in the early days of my employment in the Secretariat of the World Society in Nairobi c. 2235. Another dimension of this issue should be mentioned that supported our commitment to the educational and intellectual unification of our species.

Even in the early days of milder weather and the gradual awakening from the enforced dormancy of disaster in our world, c. 2150–2270, wise people were constantly looking back at the horrors and asking why they happened.

Words such as 'irrationality, terrorism, religious barbarism, popular stupidity' were tossed around. But even more trenchant was the view that the vast mass of humans on this earth were out of 'sync' with the knowledge that was already available in the early twenty-first century that could have allowed us to live in peace and prosperity.

A few scientists, now independent of any government support or suzerainty, dug deep into the research and evidence of humanity's relatively short sojourn into the contemporary morass. Their analysis showed that we had arrived at our contemporary dilemma bearing the heritage of two sub-species, *Homo sapiens*, the generic form of our super species, and *Homo sapiens sapiens*, the modern intellectually advanced exemplification.

I have discussed this issue of human evolution in the previous chapter of this memoir. However, the creation of a world society demanded that we get all of our chairs lined up straight so as to think about our situation as a unified biological/cultural species. How were we going to create the historical conditions for peace, prosperity and equality?

The conclaves early in the mid-twenty-second century leading toward the World Society were already of the opinion that the world community had to come together intellectually to become a truly integrated and unified species. This unification had to penetrate deeper into the biosocial potential of all peoples. The world could no longer suffer ideology and fanaticism mustered up into behavioral and institutional reality. There could no longer be room upon this earth for populations of the ignorant and the fantasists.

Those deeper sources of human emotion that we have described in the previous chapters had to be disciplined by considered, abstract thought, a cortex that could hold in abeyance the powerful mammalian drives of our evolutionary inheritance. We needed to push them off the page, if only for a moment, until the full ramification of any important question could be evaluated by our thinking brain.

Most important was the conclusion, finally certified in the first policy decisions with teeth in Geneva that the values of our racial and ethnic heritage, values of diversity and pluralism, had to be protected in any juridical and legally driven eugenic program on a worldwide stage. To the extent possible, given the ravages of the previous centuries, we wanted to preserve exemplifications of all the biogenetic inputs of humanity. Mix up the racial and national elements, fine, but always with the goal of social functionality in an ever-more abstract environment of science, creativity, talent and intellect. The motto, diversify upwards.

Look Back to Look Forward

Although it would be theoretically elegant to be able to trace the origins of such a radical development as Cro-Magnon man back to its originating moments of deviation from the earlier forms of Homo, we have to be satisfied with the available empirical evidence. This was the originating Aurignacian culture from c. 50,000 BP and in its successive movements north and west across Europe. For purposes of discussion it is enough to say that its origins in Eurasia probably preceded these tangible bits of fossil and cultural evidence by several hundreds of thousands of years, a period near the middle of the Pleistocene Ice Ages beginning c. 1.9mya.

The evidence of the rapid increase in skull carriage among a variety of other human types around the world during a broader time frame, 300,000–50,000 BP, probably should include the early Neanderthal types as recipients of this new genetics. Neanderthal endocranial capacities are voluminous. Evidence argues indirectly but persuasively that early Cro-Magnon genes were flowing outward then as in modern historical times and today.

Southeast Asia is a long distance away from the probable origins of the Cro-Magnons. At the 50,000–25,000 BP period as described by scientists we find a number of advanced erectine types. They suddenly disappeared, as with our very primitive Flores (Indonesian) hominid around c. 18,000–15,000 BP. By this time Cro-Magnon genes were probably flowing rapidly into East Asia. The successor types were hybrid peoples of Australid and Mongoloid racial ancestry, with the possibility of a small Caucasoid presence. These new genes could be metaphorically seen as a cascading set of dominos, making their way indirectly from one neighboring group to the next, east and southward, into areas previously untouched by sapiency.

In our own historical experience we have recorded the extinction of pure-blooded Tasmanian Australids in the nineteenth century. They probably had already received enough of these geographically migrating genes to pass over the sapiency line. Still, considering morphology and behavior, these were very primitive hominids. However, true to human experience everywhere, they left behind even into the twentieth and twenty-first centuries many hybridized young. The British navy did not to our knowledge set up any monuments celebrating the rescue of this valuable heritage of human genes from biological extinction. But they should have done so.

As we passed into the modern historical era, as noted earlier, the human species was evaluated by our scientists as representing two taxonomic categories. One, *Homo sapiens sapiens*, was the direct and indirect descendant of the Cro-Magnons. These historic humans absorbed the civilizational possibilities of the highly 'abstract' genetics of this new 'man on the block.' Other human ethnicities and populations now signified *Homo sapiens* included large populations within major national and geographical *Hss* enclaves. The two cohorts stood apart in the twenty-twenty-first century cerebralization of the sciences and technologies. But it was *Hss* that was central to the latter advances of humanity and the critical if painful separation of the species.

Both the League of Nations and the United Nations failed to stem the irrational violence that had erupted as a consequence of the demographic explosion made possible by plentiful fossil energy and modern scientific medicine and the consequent separation of these sub-species. I hope that the consequences have been permanently branded into our conjoined memories. The task of unification now committed to by the survivors at the beginning of the twenty-third century, has required the revolutionary intellectual and philosophical rejection of earlier ideological commitments.

Our Task

We must downsize the human numbers still weighing on the psyche of humankind. We need to stop the irretrievable destruction of much of the animal and plant life that has accompanied humankind on its journey into the present. We should continue the work of creating *Hss*, a now greatly homogenized species intellectually. This universal form of mankind will be able to come together in substantial social equality. This species of ours will still have to reshape many ancient verities so that we can face up to the facts critical to our destiny. As we have learned over the past century and a half, there still are some really difficult issues out there.

Cultural diversity, instead of material, economic, social class diversity, is the plan. The goal is peace, middle-class stability and prosperity, control over those changes in nature and society which could alter our vision of the sustainability of the human species. These programs will take place in the context of a material equality that situates our world beyond the reach of the evil that can lie within human nature.

There are no illusions as I write about the fact that it is going to be a rough several centuries if we are able to even complete these first goals. A special sub-committee of potent savants is working outside of the judicial areas where I was employed. These thinkers, talented and few in numbers, of old and young, men and women, come from diverse points of the compass. They are there for one task and one set of issues only.

Their job is to gather contemporary data from all of our constituent nationalities and their representative, locally and internationally, for indications that our various regulations and laws are resulting in individual, ethnic, national political behaviors that are the unintended negative consequences of the World Society and its Congress' policy intentions.

At the time that I left the Secretariat the sub-committee was known by its acronym, WDWDW, 'What Did We Do Wrong?'

· 11 ·
A WORK IN PROGRESS

How Much Unification?

In my own time science has come together to guide us on the issue of population stability and intellectual functionality in an increasingly abstract world of scientific technologies. Today (2284) no ideological or religious crusades disrupt clarity of thought with regard to the major issues of human concern.

The unification process produced the next question, what should the language of humankind be? Too much Tower of Babel in the world many of our judges thought. Language differences eventually lead to the division of the world into competing nationalistic ideologies. Rome existed for over a thousand years with one language. Greek was then surely a powerful and universal intellectual ancillary language.

During the Geneva period English was the language of the World Society. Neither Chinese nor Spanish could compete with English as the language of scholarship, capital, and science. As the international classic language, English does not preclude the existence of other national and ethnic languages close to home.

The larger question of the Nairobi period was the fear of a great material and symbolic innovation that could disrupt the relative calm and cohesiveness achieved in the previous fifty to seventy-five years. In my own time, at the end of the twenty-third century, we have achieved a critical reduction of our planet's population, from c. 5.5 billion in 2150 to c. 3.5 billion in 2284, 134 years of nature's enforced slimming according to policies agreed upon by the bulk of nationalities assenting to what nature had coerced.

The goal of the Directorate, the bureau of the Congress of the World Society, dedicated to the application of its decisions was to stabilize world population in the range of 1 to 2 billion humans by a gradual lowering of national birth rates during the twenty-fourth century. The Directorate feared a discovery or invention that would constitute a major disruptive material enhancement of energy supply or a communication breakthrough. This fear was reflective of a major medical discovery that had to be called back to Nairobi for consideration, an adjudication that I participated in. I mentioned this in my introductory notes.

How much control should the World Society have over the physical and material conditions under which our various nationalities and ethnicities live? As I will explain in the material to follow, we had all agreed—scientists and politicos/judges—that diversity in the cultural domain was essential. Such freedoms had to be governed by the lightest of universal legal controls. What about the physical/material side of human life?

The medical issue lies at the forefront of such discussions. We are now becoming one sub-species. Logic demands universal principles and laws to guide us in the nurturing and protection of all with regard to the health of the body and mind. This became especially relevant in the Geneva period when medical advances again began to appear and then propagate themselves into the population. These advances have led to an extraordinarily large number of centenarians. Naturally this was an especially vexing development for the young at a time of shrinking populations and opportunities. Here was one of the first judicial conundrums that had to be faced when I began my career in Nairobi with the World Society in 2030 CE.

The much-debated solution was to put a limit on the number and percentages of humans at or over the age of 100, most of whom needed special care and protection. The tentative solution was to give the individual nationalities a 10% leeway in the number of centenarians allowed, this to be re-calculated each decade. This number, give or take 10% of the centenarians in each nationality, would be allowed to live out their lives under care. This way, local cultural, even health, differences could be resolved without going to Nairobi with petitions. The young saw this judicial solution as at least symbolic of the concern for the oncoming generations. Traveling around old Kenya into the bush we could personally experience nature expressing its sense of justice in the instinctive behavior of African elephants.

A crucial if broadly envisioned decision of the leadership, with the approval of two thirds of our national membership, was directed at the issue of the new advances in science or technology. New discoveries (even as introduced experimentally) could have a possible impact on our environment, our lives, our use of energy, our agricultural methods, etc. Such discoveries or innovations would have to be put to the vote.

We are thinking here of the possible disruption to our lifestyle and basic balance with nature and humanity. Any such innovation could have worldwide importance in changing our material lives, inevitably our social structure. It could give one group of nationals a significant advantage over the others. Before we would allow such 'progress' to be incorporated into our world, we would have to have a clear-eyed perspective as to who would 'win' and who would 'lose.' Not the least concern was for the possible unintended worldwide consequences of such an intrusion into the institutional structure of our material human existence. Naturally, we first had an ongoing committee of the smart people carefully examining such 'innovations' and quickly reporting to the Congress.

An Exemplar

The haphazard 'progress' of the West during the past five centuries prior to the end of the twenty-first century meltdown was catastrophic. At that time there existed a level of human inequality never before experienced. This included pathologies of violent interpersonal crime, addiction, social and economic poverty, and widespread religious and secular terrorism. There was a wide gulf between the fortunate few and the functioning middle class as compared to the masses who were unemployed and dependent on charity.

The consequent and unforeseen destruction, the wiping out of centuries of progress amidst class warfare had to be sobering for those who survived. This so-called growth, truly anarchic, was actually seen as 'progress' by those self-indulgent purveyors of optimism before the collapse. This kind of worldly philosophy, a caricature of *laissez- faire*, could not be rationalized by those of our predecessors living a century or two after the era of chaos. When it was all over the devastation and cocooned paralysis of the human civilizational vision revealed the inherent fallacy of this semi-millennial historical dynamic.

The new approach of the World Society can be exemplified in an important issue that required my emergency return in 2278 to the Secretariat. It is illustrative of our ongoing legal approach.

The situation which developed at this time involved both the Transvaal, (South Africa) and Nanjiang (Yangtze) Province medical research. Both groups seemed to have come up with essentially similar medical breakthroughs. External medical analysts saw that each group's work was found worthy of international application and recognition. The vote in Nairobi of our science committee was an almost unanimous yes, to a harmonization of the research into its international application.

Congress' majority vote approved the dig into our collective bank account for the dissemination and application of this scientific/medical research and the special awards to those gifted medical researchers of both groups. To the individual scientists, we paid discrete sums of money or privilege; to the respective university and/or research laboratories which had supported this work, significant additional support; to the national communities which had sponsored these institutions and individuals, an equalizing grant, to be applied for worthy community enhancements.

The Transvaal folks felt that there was the beginning of antipathy between several ethnic groups in the provinces as a result of ancient racial and cultural animosities. Although the province was prosperous given their ecology, climate, and general educational levels, they decided to purchase a variety of new genetic inputs from various ethnic sources in order to unleash more genetic diversity and heterogeneity of talents into the community.

The Nanjiang people were still concerned with overpopulation. They divided up the grants to pursue in a more forceful, if still humanitarian, manner the reduction of the existing heavy weight of humans in the province. But they also allocated part of the grant in order to fund archaeological and historical research into the traditions of their community. They noted that the city of Nanjiang was the ancient capital of all of China. Both of these projects were approved by the General Congress of Nationalities in Nairobi. This vote was unanimous.

After four years of working on the legalities involving originating priorities and then the equality and character of the awards, I retired to our New England anonymity.

Progress and Stability

These issues are foremost in the challenges that the world community has to face as we attempt to create a civilization which matches the creativity of the ancient Greeks and East Asiatics with the stability in law of the successor Roman principiate. The ancient Hellenes were a people of genius but independent to the point of national suicide, even given their several centuries of flirtation with Macedonian imperialism. They showed their stuff in this Hellenistic phase of their historical run in terms of what they could achieve through military unification. They then also passed on their universal cultural vision onto the ancient world, from influencing the tradition steeped Israelites to reshaping the upstart war machine of Rome.

Rome itself saw the need for Greek intellectualism and its arts but also the need to create an empire that transmitted a universal legal system incorporating all ethnicities that came under the Roman thumb. However, Rome eventually suffered from a cultural arteriosclerosis rooted in its method of passing the baton of power. The ruling military elites, the cruelty, the insane need for control and hierarchy, even given the exception of a few far-seeing emperors, created a cultural stagnancy which seeped into all of their institutions. This decline stemmed from the century-after-century autocratic politico/military power transfers. In Rome's pagan decline, Christianity seeped into the veins of this imperium, and a new cultural/historical quiescence supplanted the old regimes. Even then the bloodletting was never staunched.

The Chinese historical tradition traded esthetics for material cultural dynamics. Although they were unable to stem the various invasions from the north and the west, they were able to absorb these foreign genes. The powerful centripetal cultural commitments of the elite transferred themselves time after time onto the new ruling cliques. The art, literature, fabrication of beautiful artifacts of ordinary life argue for a powerful and variant perspective on the nature of a creative civilization. Unfortunately this lack of attention to political, economic and then military dynamics allowed the West a moment in history to isolate and make servile this ancient cultural tradition.

The Chinese, however, in preserving and enhancing their biogenetic intellectual potency, were able to recover and dominate the world economically and militarily before the epochal twenty-first century collapse. Importantly, even with the weight of an enormous demographic bubble and the radical reduction in their material standards of life, China, now a unified ethnicity but still politically pluralized, has been able to manage its civilizational infrastructure and to bring its demography under control.

There is so much in human history that illustrates the surging drives of *Homo sapiens sapiens* for constant cultural renewal. First there are the symbolic needs of the mind for new innovative recreations of material and physical life. Here the logic of the universal mode of thought, *everywhere*, *always*, dominates our thinking. It remains a powerful human drive. We owe its awesome consequences to that mysterious set of genetic alterations in body and brain that began with the Cro-Magnons so many thousands of years ago.

But there is also a dimension of mind that gives rise to non-discursive cultural symbolic commitments, the plural. These are the softer modalities and forms of thought, the arts, ethnic commitments, sports, nationalism, and religion. Usually, but not always they can explode in creative élan alongside of material innovation and progress. We have to plan carefully, for they too are prone to fall off the highway of rationality.

Today thinkers in the World Society in Nairobi and other national administrative centers, as with the independent thinkers of humanity, experiment in binding together a union of these two dimensions of the human mentality. Our goal always has to be stability amidst progress. We try to avoid the tragedies and errors of the past, knowing full well the human proclivity to create new errors of judgment and policy. This may be nature's own fallacy in creating *Homo sapiens sapiens*.

One asks whether the bloated cancer called 'growth,' the irrationality and inequality that it spawned, the terrible mortalities of so many talented humans, the inevitable degeneration of the human stock, was worth this sort of 'progress.' Renewal in the human species seems to have always happened. Today we can understand the awful results of the shallowness of past leadership. So we ask a question: How many hundreds of years will it take for our experiment in this remodeling of the human race to include all of humankind in the potential glories of the symbolism of high culture, a rich and diverse civilization?

· 1 2 ·

UNIVERSAL AND PLURAL

Political Unity vs. Diversity

Our historians see it as important that we look back in order to guide ourselves forward. In the early twentieth century there were a number of large national blocs—Austro-Hungary, the British, Russian, German, and Ottoman Empires. In almost all of these cases these blocs were the product of aggressive national groups that were able to expand over a wide swath of former medieval and early modern nationalities-ethnicities. This general perspective would also include Italy, France and Spain, although Spain did not actively engage in WWI. The war's conclusion saw the disintegration of the existing monarchical order, which had been declining for a long time.

A great concatenation of national ethnicities was the result of this macro pseudo-political structure of the twentieth century and earlier. For example, Hungary was carved up, some put into the Czech Republic's Slovakian province, some parceled out to Rumania. The western Slavic parts of the Austrian Empire eventually became Yugoslavia, especially under the Soviet-Russian hegemon. By the end of the century Yugoslavia, a product of this universalist move of Soviet communism, had itself disintegrated into a series of genocidal wars between a variety of national ethnicities and religions.

In one part of the Austrian Empire, Galicia, the major city was Lemburg, which before WWI was described as having a prosperous middle class—one third German/Austrian, one third Polish, one third Jewish. In the countryside surrounding this city were ethnic Slavs/Slovaks, called by the Austrians, Ruthenians. These impoverished peasants claimed to have had ancient cultural roots in Lemburg. When the Austrians lost the war, Lemburg became part of Poland. In 1918 it was called Lwow. Twenty-five years later WWII engulfed our planet.

In 1918 the Czarist Russian Empire had shed many of its components, including Poland and the Baltic states as it became the Soviet Union. A new universalist ideology, communism, had been imported into Russia. With democratic pretensions they divided up the Czarist Empire into pseudo-independent 'Republics,' one of which was named the 'Ukraine,' a previously non-existent country.

In Russian, *Ukraine* meant 'frontier,' and the city of Kiev became the capital. This city had its roots in the medieval Viking era and had undergone much stress, invasions by Mongols, Cossacks, and Poles. The Ruthenian populations west of this republic surrounding Lwow or later Lvov, knew the city as the ancient home of the Ukrainians. The Russians had always called the Austro-Ruthenians, 'Little Russians', in those parts of the Ukraine to the west of Kiev.

Lo and behold, the universalist Soviet communist empire dissolved in Europe and Asia after 1989. It was no longer viable. The Soviet Republic of Ukraine became an established national entity with Kiev as its capital and Lviv (formerly Lvov) a major city. But it now had in addition to its Ukrainian (west Slovak) population, a very large indigenous Russian-speaking and Russian-affiliated ethnic population. Eventually the corruption and criminality of the leadership broke up in a coup. The Russian speakers throughout Ukraine now wanted to join their ancient ethnic homeland in 2014.

The West Europeans and North Americans disputed such a presumptive division. Had they not cultivated this 'democratic' political entity? What right had the Russian nation, now an autocratic non-communist oligarchy, to attempt to recapture the Russian-speaking Ukrainians? Conversely, what is a political unit if it does not have 'the consent of the governed'?

Universalism vs. the Plural

This particular historical mote was and is an example of the problem we at the Secretariat had to face when presenting our dilemmas of unification versus diversification to the Congress of Nations practically every time they met, three times per annum. Even as I write there are several national entities which feel they should be larger, geographically and demographically. They would like to obtain more wealth and power than the Congress will grant. They agitate, we educate. In the early days, the end of the twenty-second century, there were military conflicts over such issues.

Mostly they were guerrilla-like encounters. There was neither financial wherewithal nor the material to raise up the level of armies and weaponry that had existed in the late twentieth and early twenty-first centuries. However those nations which were then aligned in the developing international entente, World Society, did gather together enough military power to quiet things down. Today we will not allow any national entity, member or not, to have a military potential beyond a local police force. There are unhappy nationalities and ethnicities yet. But there is no existing ideology to challenge the rationality of our Secretariat. Thus, we still educate and exert economic pressures through our existing unity, even as a few disgruntled nationalities remain outside the adjudications of our Congress.

The universal drive of the mind to organize causal experience into law immediately denotes a universalism of thought. We want to order physical nature into a structure of laws which apply everywhere in the range of human experience. These tendencies go beyond the physical to the biological world, first given scientific solidity in the evolutionary writings of Charles Darwin. Then, of course, an even more universal view of our biological nature has arisen with the science of genetics and the enormous medical achievements that have been spun off from this theoretical network.

We think of our social and political involvements as also partaking of principles in this universe of laws. Inevitably we try to make laws within what is called the social sciences. This is difficult to accomplish, as Plato and Aristotle concluded thousands of years ago. But we still try to do this, and world government without national, religious, or ethnic oppression is our goal. Empires can be unified by force and fiat, as I have observed above. However, there are elements in our psyche which say, "this far and no further" will you forcibly bind us into a universal political or social bureaucracy under the force of law.

We must live with an element of universality as biological creatures, yet within the structure of human symbolic expression we also need localism, ethnic and cultural pluralism. Those who are endowed with the responsibility to bring humans together in material equality and equity have to negotiate the two opposing directions of our symbolic nature.

In achieving this balance we then put ourselves under the direction of the 'consent of the governed.' Our universal international structure of governance has to pass muster with the people's Congress of Nations.

In our major decision-making process, representatives of the world's citizens must vote two-thirds in the affirmative to put new laws or amendments into place. When dissident views develop, if peoples wish, for example, to secede from a given social entity, nation, city, or even geographic area, a one third vote in the Congress is necessary as well as a majority vote from the existing entity that they wish to leave. Naturally the departing group will have an ethnic or religious rationale for such secession.

Before the vote of the Congress the Secretariat will examine the structural conditions of the secessionist group: what their own constitutional framework will look like, plans for the education of their children, their capacity for financial sustenance, their defense of human rights, the geographical and ecological viability of the move as well as their relationship to other political/cultural entities. For the most part, the several secessionist movements that arose during my tenure have taken place with due consideration of the above concerns.

In the late twentieth century, the division of Czechoslovakia into two nations—the Czech and Slovak Republics—exemplifies this process. They represented two different religious orientations and a diverse cultural history going back before the Austrian hegemony. Their unification was imposed post-WWI by the great powers, similar to the creation of Yugoslavia but with quite different historical results. The Czechs wanted the Slovaks out as did the Slovaks. And they have lived in peace with each other ever since.

In this way we think we can open ourselves up as a species to the possible new cultures, philosophical or religious orientations and new ethnic visions of life. By allowing for such new movements within the compass of our universal structure of law, we open up the possibility for a two-thirds majority vote by our Congress that could establish a new life trend for our planet, an important amendment to the status quo.

To achieve even this amount of flexibility in our political-cultural environment we need to have the basic physical room to move around and reorganize what needs to be done for the constant arrival of the new. Our goal to create only the most modest demographic imprint on our planet is critical.

What History Tells Us

The openness of the planet Earth once allowed free movement of peoples to a better life, a land of milk and honey. This vision of the new and the possible is what has determined the World Society to allow for only a one-third yes vote in our Congress to allow for secession and possible movement. Naturally, there are conditions; there is no *carte blanche* right to take to the boats or the autos and drive off into another groups 'unwelcoming' lap or to sup at nature's new feral reserves in our continental parks.

By slowing the material dimension of change in our existence, industry, technology, even medicine, we believe we can make possible change a more local and less unruly diversion of our calm, peaceful life rhythms and direction over time. Passions, esthetic and cultural/ethnic, without the universal noise of the twenty-first century mass media will move communities but will not obliterate civilization.

From our study of history a pattern seems to present itself. In the great movements of technological, economic, or political symbolism the trend has always been to achieve larger unifications of mind. Empire building as a consequence of the perfection of the Roman military technique is but one example. The religious, military, and political passions of Christianity or Islam also united the world with the sword and gave us a new concept of the universe, here and beyond.

In the more recent past the Enlightenment undergirded by the Scientific Revolution and a still empty world brought to the West a powerful integrative vision of the rights of the individual. It underlined the deficits of the existing socio-political monarchical order, itself steeped in and supported by supernatural 'truths.' Soon after, Marx, communism, and socialism gave the masses a vision of unity and revolutionary justice. It supposed to create a world of economic and social equality. But what a terrible realization it brought.

And, of course, leading up to the great twenty-first century denouement was the universal vision of the sciences in every area of human experience. Conceptually this was a difficult structure of knowledge to digest, but it did hold out a utopian vision of knowledge and secular material wealth. However, it led to unlimited national military aggression. Many nations vied individually or as a group to translate this new knowledge and power into their own controls. Those who were incapable of grasping the new scientific knowledge skills used whatever they could muster to throw the more powerful nations into the dust with terrorism.

The knowledge societies' political and economic leadership attempted to retain their grip on wealth by sugar-coating the darkening reality for the impotent masses. They pretended to be blind to that distant looming cloud of reality, waning fossil energy resources amidst an uncontrolled demographic bubble. These two relatively obvious facts would eventually bring down the entire international edifice, advanced nations survived to an extent, but the underdeveloped world plunged into the chasm below. Out of the collapse of this universal, if irrational, vision of the future comes our current commitment of 'no more.' We hope our reconstruction will be a long-lived and historical achievement.

· 1 3 ·
ETHNICITY

Symbolism of *Ying* and *Yang*

We have been integrating certain basic understandings of human behavior and the underlying neurological causal structures for a long time now. There is, of course, this vast cortical accretion which produces the buzzing, blooming search for meanings in this complex interplay of physical and biological experiences. Include our self-created sociality and the buzzing and blooming becomes a cacophony of sensory inputs and symbolic outputs. There is no guarantee that the two will result in an adaptive outcome.

Below the cortex in our brain structure lies a multitude of mammalian-derived elements involving basic physical and sensory integrations of behavior as well as a flood of what might be called limbic system energies. This emotional overflow surges upward into our cortical grey matter to be organized and turned into useful social symbolic forms of meaning. These two systems can be recognized in lower mammalian forms and in our hominid ancestors. In the morphological revolution which is Hs and Hss, the derivative can be civilization or barbarism.

The emotional juices of the limbic system can certainly give tonality to our most mundane intellectual/cerebral tasks. Scientists and inventors have testified to the emotional qualities that their seemingly hard-nosed work requires. But the symbolics of the limbic system even as shaped cortically into tangible expression involve the softer forms of culture, the arts, religion, sports. Sadly these emotions show themselves harshly in such cultural institutions as patriotism, the flag, uniforms, weapons and the surge to kill.

The material world as expressed symbolically tends to lead to questions of cause and law, then equations, which speak about universal relationships. The names Copernicus, Newton, Einstein tell us about this thrust of mind.

The engineering of an airplane may involve certain cultural stylistic choices but the mechanics and design demand a universal authenticity in order for the plane to fly. Kuala Lumpur state-designed and -manufactured aircraft obey the same rules of physics and engineering as do those manufactured in the nation of Texas. This universality of symbolic meaning has been called "discursive" symbolism. This is what science is about. Theories of infection, the cures, medicines, and finally, the hospitals don't vary very much from one ethnicity to the next. In a world that has come together as the World Society is constructing it, esoteric diseases in distant lands will not be communicated. They may occur in animals and then in humans. The hospital cure will attempt to be universal in nature.

We cannot say this about music, the fine arts, even sport. Plurality is the word for so-called nondiscursive symbolism. However the two realms can interact in the arts. A symphony may be written in a universal pattern of orthographic notation. The instruments and musicians that play the music may even be international in origin and training. But the compositions that emanate from the minds of the composers will tell us that its origins were in, say, Italy or Poland. To the ear the music of nations will 'taste' differently. In fact, we have learned to treasure diversity, pluralism, in a wide range of human interests and endeavors. The dynamic of expression in this world of the non-discursive or the ethnic is a force which no polity can ignore. The key to understanding ethnicity is to observe its role in human history.

Definition of a Nation

Take Rome and Latin as an example. Remember how Latin expanded into the Roman provinces of the West. Its amplitude was an outcome of Roman political, cultural and military power. The western indigenous residents of this geography were Celts; the Gothic immigrations from the east were Germanic.

When Rome in the West disintegrated, a wide variety of so-called Romance languages arose. These were distilled down to a few as the powerful locales overcame the weaker. Gradually the national languages as we have come to know them arose during the defining of a nation. But even today we hear the accents, the residual patois of the ancient residents of these areas. And with the disintegration of the larger power political units, they have gradually returned to reclaim their symbolic place in the sun.

Whenever a great universal movement loses its innovative symbolic dynamic, inevitably pluralistic elements begin to gain energy, usually arising from the above-noted linguistic or religious self-recognition by the indigenous populations. For example, I cited in the previous section what had occurred in the breakup of the post-WW1 era. The rise of communism and fascism with their universalist ideological claims seemed to launch a new era of submerged ethnicity. War and philosophical irrationality dissolved this movement. New political entities came into being, largely reclaiming the underlying ethnic, religious, linguistic symbolism that self-defined the populace.

Such a break-up happened to the United States at the end of the twenty-first, beginning of the twenty-second century when the weakened and incompetent central government in Washington, DC could no longer gain the support of the various regions. As a result of the federal bankruptcy a number of the states broke away; others joined together as semi-secessionist political and economic entities. Because the most economically self-sufficient entities were the ones that broke away, Washington, DC could not muster a powerful enough defense of its authority.

When the central meaning of a nation dies, the suburbs try to survive. New ethnicities emerge; the triage of the uneducated and unemployed whittle down the respective populations. The older and hidden diversity of the North American sub-continent began to return as new regional economic energies reshaped these now-autonomous states.

Language: Two Functions

We *Homo sapiens* are unspecialized in our adaptive profile. We can't compete with other animals in terms of hearing, seeing, running, climbing. We need social help and tools to hunt and eat. Indeed, our brain helps in thinking out pathways for survival. This one specialization, language, tells it all. Even the Neanderthals, Cro-Magnon's competitors in Eurasia, may not have been able to articulate vowel sounds. Did that deficiency lead to their extinction? We may never know.

Language for discursive communication, to speak about the external world in secular informative fashion, is of course the major direction of linguistic discourse. Even though language seems to act as the pivot for the cerebral release of human adaptive behavior, we have tended to neglect poetry, myth, and song. These so-called nondiscursive symbolic linguistic expressions seem to derive from a wholly different set of neurological functions. They certainly reflect a different reality of human language functioning.

Language can bring us together to solve tough intellectual scientific problems. But this Tower of Babel multiplicity of 'tongues' can also in deadly fashion separate human from human. Why should this be? Clearly language is communication, a critical element in the social glue that allowed us as unspecialized animals to bind together. Why is it that we can easily interbreed with those odd speakers at the other end of the valley? Just a few generations ago we could easily understand each other. But with little contact since, in but a few generations we have become strangers.

Language differences can create deep cultural valleys. But for the most part we can translate the disparate symbolic statements of one language into another. The fact that we live upon this earth together must create a similar set of symbolic images within our cortex so that we can function.

To manage this ever-ongoing diversity of speech, humans in their attempts to attain universal power have inaugurated universal languages for the ordering of society. Even today we need to bring our world together politically and socially in an agreed-upon universal spoken and written language. But we also feel that it is necessary for humans to be able to return to their homeland and the intimate richness of the mother tongue.

Writing and Speaking

The invention of writing by the Sumerians of Mesopotamia, some 5300 years ago, to communicate various discursive economic relationships was quickly followed by the extension of the written word to record the myths, tales, histories of the various communities. Soon came a written literature; there was already a rich oral linguistic tradition—song, poetry, storytelling—to widen this universal biological facet of our nature into the various domains of symbolic expression. Eventually the concept of the written word would lead to the concept of the mathematical theorem, then logic, resulting in the broadening of discursive knowledge, inevitably, to the universal imperative of law and science.

Whereas the Sumerians and the later Mesopotamians used a syllabic form of writing, the later adaptation by the Chinese Mandarin elite was a quite abstract and intellectual rendering of the written symbol. Chinese orthography is called logographic. It utilizes graphemes as symbolic tools, a purely semantic creation without mandated phonetic equivalents. In contrast the Phoenicians and then the Greeks attempted to create an alphabet closely aligned with the spoken phonemic structure. The earlier Mesopotamian and Chinese forms of writing were limited to a restricted class of intellectual elites. The later alphabetic forms were aimed at more general populist usages.

Educated Chinese throughout the realm could read the communications from the center. Because the writing was ideographic they could automatically translate the written symbols into the oral version, the local dialect. In traditional China the spoken language between the various valleys and provinces was often mutually incomprehensible. This was (and is) not the case with the written Mandarin form. The fluidity of the non-discursive dimensions of the human mind allowed the spoken version to take on its own dynamic of change while the universal discursive written form remained relatively fixed.

The universal Catholic Church held onto its own Latin exclusivity, but various nationalities after the fall of the Roman Empire were able to go their own way as contrasted to the Chinese dynamic. Thus when the oral forms of the evolving Latin were transcribed into writing, they now deviated sharply from written and spoken Latin.

Thus we see from the most emotional, often hateful imprecation of the limbic system to the most lofty philosophical essays, language encompasses a vast span of psychological and symbolic intent latent in the human brain. As we come together as a World Society we need to balance out these two seemingly opposed directions of thought as expressed in language

Importance of Ethnicity

One thing I believe we have learned. The material dimension of life, tools, technology, machines are not as crucial to a true civilization as heretofore thought. Throughout the world today, as I write in 2284, there are hundreds of monstrous architectural relics standing, grizzled empty shells of a once-exuberant world. The nationalities wherein these buildings cast their daylight shadows are yet without the resources to take them down.

In those five hundred plus years of material expansion in the developed world we saw growth and the accumulation of wealth that exceeded all expectations. The world of junk consumerism was born but accompanied by real scientific and technological advance. And of course, up until the last century or so before the fall, people created historic art, music, and the nondiscursive intellectual abstract activity of an excited world.

Predictably, all wonderfully irrational adventures end; here they were burned into the history books by the terrible wars and genocides that came before the energy collapse, the change in climate, and the great fall into chaos. Nowhere had there been a dream of a factual democratic, egalitarian civilizational future! In our own planning the bloat of demographics and economic and material expansion are under sharp scrutiny by the nationalities of the World Society and its Secretariat. The truth is, human actions with regard to the external realities of life have their sad consequences, too often unintended.

Our focus then is on the deepening and disciplining of our natural tendencies toward ethnic diversification and separation. First we realize that all nondiscursive symbolic creativity takes on a local ethnic coloration from musical and dance styles to architecture and sports competitions. Ethnicity is our guardian against the grey homogeneity of political absorption and the commercialization of our lives by international cartels.

Being European educated, I can speak about our research into semi-independent nationalities and their role in history. Other colleagues and previous researchers have dug into the traditions of the Americas, Asia and Africa to come up with interesting historical narratives of the survival of independent states, often city-states that created an internal ethnic purview of life parallel with and in connection with sister states. A consensus has developed over the past several decades that we must focus on the city-state or even rural mostly agricultural communities as exemplifications of the good creative life.

The Small Was Beautiful

Exemplars from history include the early Sumerian cities from 4000 BCE—Eridu, Ur, Larsa, Uruk, Kish, Lagash—all almost forgotten now but at one time in intense competition to be the embodiment of Sumerian civilization.

Sadly this competition eventually led to resentment, wars, finally autocratic monarchical supernaturalism dominating these towns with a brutal undertow. But these independent towns were once centers of internally motivated ambition to be the best.

In the earliest period of their advance they adopted the more ancient democratic tribal traditions of the Cro-Magnons by voting the wisest and strongest individuals into leadership positions to be 'the big man.' They competed in their institutional religious endeavors, in architecture, the arts, in the economics of agriculture, then international trade, levels of literacy and produced a rich mythological literature. They even taught the Egyptians about engineering and building the pyramids.

Similarly, the Hellenic city-states were rich in esthetic and philosophical/scientific innovation, internal patriotism, even a true constitutional democratic structure of urban decision making. Athens, Sparta, Thebes, Corinth, Megara competed to be the best (*arete'*) in their love of knowledge, beauty, individual accomplishment, the Olympics and unfortunately, heroism on the battlefield.

In the early modern period the European city-states germinated individual liberty in the face of the overbearing orthodoxies of Papal Rome, even Calvinist Geneva, and especially the feudal agricultural aristocracy defending their ancient privileges. Still, great intellectual achievements and art were nurtured in these towns. Earlier in the High Middle Ages even cathedral building had become a proud community effort. Rome's embrace was distant. God's grace, however, shone down from above.

In Italy the towns spoke different dialects yet were still understood by their neighbors. The distance from Denmark and Flanders to Italy and Spain did not daunt the artists and intellectuals who wandered freely over fluid boundaries. A European-wide culture of unity in difference (*E Pluribus Unum*) was created. It is interesting to note and puzzle over. Even as the city-states were merged into great nations in the eighteenth and nineteenth centuries, the level of creativity maintained itself until the ideological madness of the twentieth century poisoned the well.

Future of Ethnicity

The World Society wishes to create an environment such as existed in the heyday of this rich intellectual ferment. Today we humans can speak two languages, English and our own; we are citizens of the world with a patriotic sense of our uniqueness. Naturally there will be limits to the 'uniqueness' of any particular ethnic variant that develops. On the other hand we have in place a way out for any dissident group that can carry its own weight financially and constitutionally.

As we lower our demographic profile, becoming increasingly productive in the essentials of modern middle-class life, we will gain the surpluses necessary to cautiously expand our material/physical imprint. Before these conditions become a worldwide reality we will have to nurture the internal political machinery in the existing nationalities and in Nairobi. We will always have to carefully scrutinize and subject any change to rational deliberation before we go whole hog.

Meanwhile, energies in the arts and other cultural or civic enterprises can foster the creative symbolic needs of the young and adventurous. There is much knowledge yet to be recovered and then used to push the boundaries of philosophy and science. Energy efficiency and surplus are still only on the horizon. We are building the internal societal disciplines and gradually funding the research to begin this great structural enterprise. Most important, we will never politically ride roughshod over our ethnic communities as was done from the mid-eighteenth century for ephemeral short-term gains in power for the few.

A very important part of human nature is rooted in the face-to-face community. We humans have a universal drive rooted in our mentality both to know and to do. That is why we need worldwide political and legal governance to guide us forward without conflict and inequality. But more importantly, there is the basic need for the freedom of communities of the like-minded to independently frame their futures, nourishing the unique tastes and qualitatively plural visions of those aspects of social life that we call ethnicity.

· 1 4 ·
OUR DEMOCRACY

Is Democracy Possible?

From the very beginning of the dream of a world government and a democratic society, there have been plenty of naysayers, even among our own people. The record is not good. The stories about the first urban civilization in Sumeria are distant and vague. A democratic polity may have taken place when they were still little more than tribal ethnic communities in the possible model of the Cro-Magnon tribes of the North from which they probably descended after the waning of the ice flows.

By the time literacy was established they were clearly urban with a complex economy. The scholars were relatively isolated from the people. Reading and writing now were skills handed down from family to family a secretive knowledge not yet available to the folks on the street. You cannot have democracy when the majority of the population does not have the knowledge to say no to the leadership. And there were complex decisions to be made in a city of specialized vocations, separated from the ruling classes such as military, priests or monarchs.

Thousands of years later the Hellenic population was definitely literate in their centurylong heyday of democratic procedures. The dramatists and

Sophists gave them something to read. The general assembly of the polis successfully functioned as long as the Polis was economically on the rise and militarily successful. By the time the Athenians lost the war to the Sparta, Pericles, their visionary inspirer was long gone. The ordinary folks decided to indict and try their hoplite hero, Socrates, a plebian from their own working class. He was tried for treason having publicly laid out his dialectical teachings, supposedly to the rich guys friendly to Sparta.

Socrates decided not to run away and thus his death sentence was carried out. He had obeyed the patriotic injunction of the city, then going down to defeat, to have two children at a late age for the possible renewal of the polis. We know these facts of democratic decline because his brilliant aristocratic student, Plato, wrote it all down in beautiful powerful prose. The democracy sputtered along for a few more years until Philip the Macedonian finally defeated Athens at Chaeronea. Both Plato and his rustic student, Aristotle, threw up their hands at the thought of democracy, seeing the Hellenic masses as prone to manipulation and ignorant hysteria.

The English democracy of the eighteenth century endured for several centuries, but the economic and aristocratic elites always held the reins firmly in hand, doling out to the multiplying masses the minimum in ill-paid employment and education. The will of the people was hardly being expressed.

In the United States we had another heroic beginning equal to that of the Athenians in vision. It was put into place by a group of very talented men, squires, of means but of a relatively homogeneous ethnic background. But this nation was created on a blank slate. It was constructed on a continent of vast natural riches untouched by human hands except for those of the unfortunate Amerindians, who were on a different cultural and historical level. In addition, the settlers of European heritage early committed to and had to subsequently face up to their introduction of black African slavery.

From our twenty-third century perspective the American citizenry at first were able to keep up with the leadership in terms of ongoing knowledge of affairs. However, after a terrible civil war during which slavery was brought to an end, the democratic direction of this nation was definitely in the hands of an economic and political oligarchy by the late nineteenth century. By the twenty-first century it was only a matter of time before the existing social divisions would tear this nation apart. The masses had become subject to powerful leadership groups, not comprehending in any way the disaster that the well-ensconced governmentally supported elites were leading them toward.

The World Society Governance

It was a great privilege for me to work as a member of the Secretariat Judicial Board. It was one of a number of committees of the Congress of the World Society, a locally elected international governing body. As long as no member or group of members of the Congress raised objections to the various study boards, we were reappointed (or not) every five years. The Secretariat itself had an internally elected governing committee to self-examine its workings, ethically and competently.

All of the decisions of the Secretariat Judicial Board, although constituting what in earlier times could be called a world court, needed ratification by the majority vote of the Congress. Members of the Congress were elected from their home nationalities and ethnicities for a nonrenewable period of ten years. In the home nationalities there existed a variety of governance structures, usually with strong judicial and legislative privileges, the executive with nominal executive obligations. The latter were also obligated to supervise the laws and decisions of the local legislative and judicial institutions of governance as well as the more universal international issues that came before the Congress of the World Society.

Member nationalities and ethnicities were admitted to the World Society by application and majority approval by the Congress. Each political entity had to meet certain democratic criteria of governance, fair elections, a judiciary having a longer term in office than the legislative or executive branches. Some nations were bicameral; some judicial structures were subject to the plebiscite, others appointed by the elected legislative bodies. With regard to the internal political organization of the member societies, much variance was allowed. It was only when substantive internal objections from the publics of the nationalities were received by the Secretariat that international organizational scrutiny was applied.

Because the primary objective in the early stages of this world government both in Geneva and then Nairobi was to get a handle on the demographic issue, the nations and ethnicities admitted had to commit to a population that was not out of kilter with other less populated or smaller geographic imprints. There were a number of close calls here as large political entities had to be broken up, the division process bringing to the fore antagonisms which occasionally were on the verge of outright war.

Even in the earliest days of international negotiations, the bulk of the nations who wanted to create a new order of life maintained the basic political vision. No nation or ethnicity was to be organized as a super-state in population, geography, natural resources. The Congress was implacable in its resistance to large militias that exceeded the actions of local policing. The World Society alone had the obligation to defend the peace. The World Society military board early on established high-technology bases in various geographic areas and nationalities around the world to ensure the peace, not allowing any significant armaments in member nations.

When local politicians had fulfilled their maximum service to their respective nationalities, they could then run for office for the Congress of the World Society. Service in the elected Congress had its term limit. As a member of one of the committees of the Secretariat which had only informal limits, I served for forty years. However, I was never elected to office by one of the nationalities. All member nationalities/ethnicities had to put term limits on all their governing officials—judicial, legislative, and executive. In general there was room in our organizational structure for the examination of all issues of democracy.

A two-thirds vote of the members of the Congress could bring any issue not already presented by the Secretariat to the Congress to the floor. Because these members were elected from their home constituencies, the Congress had been empowered to alter the basic constitutional structure of the World Society, that is, to make an amendment to its basic historical functioning. There has been no problem in achieving such changes by vote as we are still in a fluid political and economic state of world affairs. The model of governance here is closer to the moving constitutional vision as developed in Britain in the eighteenth and nineteenth centuries than the established written constitution seen in the old United States and as existent in other countries.

One needs to add that the member states may or may not have their own written originating documents. Also, unlike the failed United Nations of the twentieth and twenty-first centuries, the World Society has no Security Council. The reason for this is simple. We do not believe in mega-power nations. To become a member of and send delegates to our Congress, the applying nation must be equable in population, geographical extent, natural resource availability. It is our conjoint responsibility, national and international, to oversee their intellectual and educational progress, the internal democratic functioning of the society, and the general tonality of its contribution to the international state of humankind.

Democratic Vision

We saw ourselves as an innovative political body. Unlike the *tabula rasa* of the constitutional founders of the United States after their liberation from the English monarchy, our reconstruction of a world order was built on worldwide ruin. We had to cleanse the political landscape of a wide variety of embedded political and ideological microbes. These microbes were the source of the human wreckage, the stagnancy of abandonment and hopelessness which had paralyzed humankind for so long.

When our originators were first gathering to pick up the pieces it was with much hesitation. Many of these thinkers were concerned about the outcome of such a confrontation with so-called embedded truths. The new ideas were truly radical, if not revolutionary, perspectives on human nature and society. Thus, except for the issues of equality in our international human intellectual capital and long-term demographic weight reduction, all of our structural political and economic solutions were tentative, awaiting the feedback of ideas and events. In effect we have followed common law traditions.

Also, as we are believers in the democratic vision of human social existence, it was clear that no democratic political and social experiment could exist as an island in a sea of autocracy. And thus the great new world that our philosophers and scientists wanted to try to build had to be a worldwide reality. It is thus that the intelligence factor loomed large in the desire to create conditions of human equality and then liberty. The citizenry of the world had to be able to think through these issues using reason, not through their 'guts.'

Naturally humanity's historical experiments in democratic polity came under careful social science scrutiny. We had to make the transition from the naïve and small-scale tribal traditions which still existed in the nineteenth century. These were presumed to be at the heart of what seemed to be a stable socio-cultural element even in the long-term Cro-Magnon preliterate civilization so many thousands of years ago. We had to add to the historical equation factors of relative economic scarcity and climate instability. These factors were important in studying the viability of democratic solutions of the past. As the fine arts loomed ever larger in our vision of the democratic life it became clear the original florescence of art within the Upper Paleolithic Cro-Magnon tradition could not have been so long lasting had these people not lived in a stable socio-political and economic environment.

So too, understanding was needed with respect to the supposed economic and political stability of the Sumerian cities in the rich swampy ecology at the junction of the two great Mesopotamian rivers. These immigrants over a period of several thousand years constructed out of these swamps, first, an agriculture of surplus and then an urban environment which led to skill specialization. Here again the probability is that the age-old tradition of the leadership of the elders, elected by their peers to guide these newly arrived wanderers over some difficult challenges, both military and economic, created the 'big man' executive. This political structure seems to have united in his person both the military and religious elements in the ethnicity of the various multiplying communities in that delta region.

Eventually the alliances of townsmen and rural agriculturalists became too big to handle. Conflicts eventually created not only military/religious tyrants but invited the upstream Semites to overwhelm a once-democratic culture that had fallen into decay. Naturally we attribute this fall from grace to their lack of knowledge of the greater universe of causes and ideas.

As a result of such studies we have become well aware of our own limited knowledge to venture too far afield in our constructions, material and conceptual.

Still, the greatest loss for the democratic ideal occurred in Hellenic Greece. Several thousands of years after the first urban experiments, the Greeks gleaned much from their own historical perspectives. And although the democracy of Athens shone brightly for a number of generations, including a great patriotic war against a far more numerous and powerful invader from the East, they could not hold together as Hellenes—too much sudden wealth, too many independent city states contending to harness the power of this wealth, bloodletting among Greeks that matched the tragedies of the twentieth and twenty-first centuries.

But like the Sumerians they gave human history so much in the way of a secular philosophical articulation of the free human individual and the arts and sciences that the minds of free humans could produce that, all is forgiven. The Hellenes were probably the most intelligent *ethne* that humanity has produced in the context of a highly developed urban way of life.

That is why when we look back to the American revolutionary experience we note the unlimited empty richness of this geography and the tragically underdeveloped Amerindians who were routed from their primeval landscape. These revolutionists and constitutionalists were already harking back to ideas produced in Europe by both French and English thinkers. They were aware of the now-limited monarchy and the parliamentary system that the English had established to guide their own imperial visions of power and wealth.

Sad to say, Americans quickly adopted black African slavery as a method of economic advancement. In the short run this diversity of peoples, even when slavery was dissolved after much bloodletting, held true as a new democratic vision. But this diversity of background and historical development led to policies that eventually turned this land, once one of 'milk and honey' into an impoverished sub-continent of political and economic chaos.

Throughout the world of the twenty-first century democracy had commanded a surface nod of recognition. The powerful political classes learned how to debase the intellectual level of supposedly democratic decision making. The key was the propagandistic use of the pseudo-plebiscite now to put it over on the media-mesmerized ignorance of the masses. Of course it was only a veneer of the democratic principles that the original American leadership had presented in the Enlightened eighteenth century. True, vast industrial growth and population expansion played a role in destroying the democratic ethos. Certainly, the knowledge base of the so-called hard sciences did not spill over into socio-political understanding.

A tyrannical government like communist China could lure its highly intelligent fellow ethnics into enormous technological and economic growth and progress, all the while imprisoning their deeper human nature in a solitary confinement of propaganda and police oppression. Oh, yes, the ruling autocrats made themselves very wealthy before this burst of prosperity fell apart. Fortunately for us, China has now disintegrated into an interesting diversity of political/cultural entities, still struggling with the effects of demographic indigestion.

· 15 ·
WHY DEMOCRACY?

Perspective

One small historical fact seemed to tip the balance of our predecessors thinking when it came to planning the way forward for the World Society. The Enlightenment had swirled over Europe like a cool breeze of excited hope. The American Revolution itself seemed to hearken to a day without monarchs or police lockups for those impudent enough to speak their piece. Joseph of the Hapsburgs c. 1775, son of the tough Maria Teresa and brother of the soon-to-be headless Marie Antoinette, was one monarch who decided to go with the new breezes.

Emperor Joseph II inaugurated a wide series of anti-clerical, anti-aristocratic agricultural and economic reforms. He saw that the days of absolute monarchy the likes of which his mother advocated could no longer withstand the avalanche of science and knowledge coursing through the Western world. Unfortunately he died young; the French Revolution soon bloodied this dying institution of monarchy. His replacement, Francis, still had the power to erase practically all the reforms. There were no institutional bars to such a reactionary ploy. A bit more than a century later the Hapsburg Empire no longer existed. Millions of the world's finest young men were lost in the conflagration of WWI, marking the end of monarchical power but introducing us to a greater ideological tyranny.

Indeed non-democratic leadership can, under unusual circumstances, carry a people forward more quickly than through the arguments, discussions, often political paralysis of democratic decision making, with their parties, cliques, class warfare advocates. Historians point to Joseph Stalin's five-year plans for Russian industrialization c. 1930s or his Czarist predecessors in the last decade of the nineteenth century and their rush first to empowerment, then to world war and revolution.

Then there was Deng, who after the death of the viral tyrant Mao in 1976, attempted to combine autocracy with looser economic reins. Indeed, it catapulted the Chinese people forward, but as with Stalin's absolutist pitch, his genocides and disastrous wars and then continuing autocracy, the consequence for the Chinese nation by the mid-twenty-first century was a vast depression, an industrial machine that was an energy-deprived hulk exporting zero to the impoverished around the world. This depression finally saw the breakup of this once-glorious Han culture. Today the remnants of this culture will probably exist for the long run, a heterogeneous, linguistically separate group of states now attempting to shed the burden of masses of poverty-stricken dependent urbanites.

Our thinking is this. The demographic relationship of humans to resources is now moving toward the long-term stability of the species. Now that we are on our way to creating one species, never again to be divided by the old claims of racial, ethnic or social class prejudice and discrimination—in short, oppression by one defined group upon another—we could be on the way toward smoothing out ancient grudges and human envy. True, these are only a few of the social elements which have torn apart the democratic experiments of the past. But we are now approaching a point of equilibrium. We can continue to build upon the future.

Another factor favors our hesitant experiment. The lines of communication, the technology which in many ways destroyed the old nationalistic, ideological system of the twenty-first century, have given us the same, if not a better opportunity to know each other, to speak to each other in a language that we can all share. The English language while maintaining the ancient orthography, uses phonemic accents very different than those of several hundred years ago. We laugh when our youngsters have fun in resuscitating and listening to the old tapes and transcripts. Our descendants will do the same to us. The point is, even this old universal language is new.

The classical Greeks had the paradigm conditions for a sophisticated vision of the democratic life. They were technologically and culturally advanced considering the state of civilization at that time. The citizenry were probably more attuned to intellectual issues than was the world in the late twenty-first century when everything fell apart in stupidity and corruption. Pericles in Athens presented a complete democratic vision after the Athenians had built on their surprising and exhilarating defeat of the Persians and the latter's Greek allies.

This vision, however, fell apart with the sudden death of Pericles at the start of the internecine Peloponnesian war with Sparta and its allies. He made the error of inviting within the walls of Athens all Greek citizens in the outlying Attic peninsula. These Greeks brought with them a plague that wiped out a third of the population and Pericles himself. But the adrenalin of war was in the air; it could not be damped down.

Certainly Pericles was responsible for an imperialist thrust at his vaunted spiritualization of Athens with its great religious/esthetic construction on the Acropolis. The war was eventually lost for Athens. Socrates, their prickly intellectual goad, was condemned to death for asking too many ethical questions about the nature of their social vision. Plato and Aristotle, intellectual students and descendants of this quest for righteous understanding, were skeptical of the possibility of democracy, given the manner in which the demos had voted and acted irrationally when in the grip of their war power to rule.

No, the golden souls themselves could not lead the state, wrote Plato. Even these, so carefully chosen, could not predict the proper passages of the stars and planets so as to order the workings of the society. Aristotle in lectures at his Lyceum saw all human behavior and its social realizations as only probable knowledge, in contradistinction to the 'hard' sciences. Perhaps a limited monarchy might work, he hypothesized. But the question of how much monarchy and how should it be limited went and goes unanswered.

The wisdom of the Greeks in confronting human nature, here their own ethnic flesh and blood, has to be a perennial exemplar of the societal message, 'go slow, folks.' We too hesitate even in our enthusiasm for a revolutionary turn away from the philosophical misunderstandings of the final version of the egalitarian, pseudo-growth and technocratic models of the twentieth and twenty-first centuries. What with the constant bloodletting in these centuries, too often destructive of their most educated classes, we must constantly bear in mind the possibilities of our own errors, now and into the future. After all the so-called democratic leadership groups, even the totalitarian dominators were leaders, supposedly the smartest people of their eras.

Democracy's Rationale

There are, regretfully, only a few moments in human history when a democratic polity is able to enact a revolutionary turn. More often such sharp moves of reconstruction are taken on by monarchs or tyrants. In the early stages of such 'reform' there may be a temporary exhilaration. In the longer run, that momentary success brings forth new dynamics which most often cannot be absorbed into the understanding provided by those earlier positive hints. These unplanned-for consequences are usually met with the sword of power. The original brokers want to keep it all, even when their moment of achievement has long passed. And always, the concentration camps will fill up.

The democratic process has to be slow and conservative. It takes time to discuss, analyze, persuade, to look for those possible unintended consequences that new directions always throw up as a surprise! And, of course, part of this always slow deliberative method is keeping power pluralized and out of the hands of the potential autocrat.

At the core of the rationale for democracy is the reality that progress has to involve the will of the governed. Eventually, deviations from this rule end up in conflict and ruination. Democracy throws out its rays of policy determinant to all the people, demanding to receive back their educated perspective. It is slow, but it makes fewer mistakes, because it derives from this will of the governed, their wisest choices. It cannot withstand terribly dynamic moments of economic crises, nature's revenge, nor the bloody wars which destroyed both the Greeks and our modern-day pretenders, who are, in reality, purveyors of irrationality. Democracy survives through the will of the intelligently governed who require active participators in the process of choosing/deciding.

At some point, perhaps a hundred years down the line, when our demographic task is completed and behind us, when the intellectual harmonization of the species seems to depend on talent, practical efficiency and the various disciplines of skilled labor, the elected leadership will then demand a more discursive and permanent constitution emblematic of our past and a commitment to the future.

As it is we have been able to gather together under this one political umbrella the unity of the nationals and ethnics of our world for barely a century and a half. It hasn't been easy or without modest conflicts. The urgency of the basic democratically agreed-upon tasks has absorbed us, along with the careful distribution of natural resources, the re-linking of humanity in communication and then in modest transport. We are talking with each other. Representation within the various states is fluid, always, new guys and gals on the block.

In our two international capitals, Geneva first and now Nairobi, the Secretariat and its various committees are more stable. They say our world needs experienced, long-serving managers and facilitators. I was one of the longer-serving members. But we in the Secretariat are always subject to immediate removal by a majority vote of the elected Congress of the World Society.

How will our democratic functioning endure when the basic commitments are largely completed and in place? What new human intellectual and emotional forces will arrive on the scene? We can't predict. It could be a desire for either a more hierarchical or a more fragmented, centrifugal, decentralized structure. It could demand more freedom for materialism, economic competition and faster innovation, even more national enlargements and growth. We can't predict now.

Our only hope is that the general uniformity of high intelligence and higher education will lead to the cortex winning out over the emotional system in the decision-making process when faced with issues of public political and economic concern. We are bereft of instinct to help us decide what to do. Our hope is that the democratic resiliency of intellect and rationality will take us safely over the shoals.

· 16 ·
THE DEMOCRATIC LIFE: CRITERIA

Democracy Sustained

The conditions for democracy as elaborated below, were in the forefront of the early planners' thinking. The clouds from the political disaster of the recent past were still a haunting presence. From this debris a new beginning, a new page in human history, had to be constructed. The democratic social and political experiment had to be given another chance to function in an advanced scientific/technological global civilization.

Rational Decisions

Nature created *Homo sapiens sapiens*. We humans are an example of the dynamics on our planet that have precipitated constant change and challenge. At every moment in the history of our species we have had to face up to the challenges that nature has thrown at us. This demands that our brain, our intelligence can meet these changes so that our species can survive.

The human species is a complex composite of behaviors experienced and potential. It has created complex civilizations, then alterations unlimited, disasters, accomplishments. We do not know the limits of human self-innovation in conjunction with nature's own dynamics. Thus we need every bit of our rational brain power to meet the surprises that our own human nature will throw at us.

In short, without the instinctual guidance that nature has given other forms of life to either meet these changes in nature or most often to fail and become extinct, we have to rely on our brain power, our rationality to look not merely at the consequential events caused by both nature and man, but most importantly the causes and the underlying structures of our reality.

A: Democracy and Human Intelligence

The dynamic of human civilization tells us that there will always be ever-greater complexity in human symbolic interactions. The knowledge required to function, especially with the increase in the mechanization and robotics of labor, requires ever more abstract knowledge as part of our education. The implicit powers of the original Cro-Magnon, *Hss* mind, is clearly realizing itself in the world we are trying to create. This is a civilization denuded of religious irrationalism, ideological totalitarianism and all the other venal pathologies that brought down that pseudo-civilization one hundred years after the late twenty-first century.

Democracy requires decision making by the large majority of the citizenry, local, national, international. To make rational decisions they have to know as much about the issues in question as their political representatives. Citizens have to become free economic atoms in the society, working for themselves and their fellow citizens, not dependent on the government job or dole.

Sharply divided classes, economic and social, racial or ethnic, reflect the failure of education to equalize. In the old times such divisions could be attributed to the power of one group—racial, ethnic, even a political super-class—to control and discriminate. In our construction of society within the community ethnic and racial differences are disappearing rapidly, and a community-wide high bio-social intelligence is being carefully nurtured. If we succeed such divisions should not occur. The key for each community is to nurture its potential for high intelligence and high education. This will become the major protective device of the democratic process of decision making.

B: Knowledge and Education

The need for both local and national educational commitments as well as a dominant international focus is today felt by an increasingly interconnected world, even beyond what was achieved before the great fall. Our goal should be for the people of the world to have the same basic repertoire of knowledge skills required for this new international vision. We suggest an international version of the old liberal arts education that was offered in the Western nations.

Now that we are well on the way toward eliminating the old racial differences, experiences and prejudices, we have to absorb the reality that the transferred and transported genes do not carry with them the imprint of history. Although we encourage the movement of peoples as tourists or immigrants to interact and fructify existing communities, gene transfers contribute an ethnic and psychological history that is implicit in the children born of such redistributions.

Thus our educational focus has to be the re-distillation of the knowledge and experience we have gleaned from the history of nations once called France, Afghanistan, Viet Nam, or Costa Rica. This reinterpretation of the past will now arrive educationally without the incubus of personal ethnic or national memories. The burdens of the past which once weighed down so many nations when faced with the 'new' will not affect us; there is now a true *tabula rasa* for social planning.

What we have learned from these varied historical experiences of peoples and nations should constitute a good part of the liberal arts education that all citizens of the World Society will receive, but also, the scientific knowledge, no matter from where it has and will emanate, should constitute the contemporary knowledge base of all world citizens. Then, we can localize, as once did Texas, a state in the old United States of America, which required of its citizens courses in the history of the state. This is fine; we need to harmonize local, national and ethnic allegiances with a commitment to the stability and adaptability of all the people of the Earth.

Thus, when the local national assemblies convene to legislate and then carry forth this legislation, the voting members of the local democracy will know that they have gained enough knowledge to have their say. This applies equally to the members of the Congress of the World Society. In terms of knowledge and the practical truths which will derive from this knowledge base, the democratic processes initiated in legislation will have the sanction of the citizenry.

C: Ethnic Homogeneity

Indeed the transfer of genetic material from one geographical/national location does not necessarily carry with it the great ethnic imprint that race once did. Yet, people living together, marrying, having children, have over the generations acquired a genetic physical, personality imprint congruent with their neighbors down the block. In the old days, the neighbors were often cousins.

However, as sociologists once believed with religious intensity, we were infinitely malleable in our behavior because humans lacked an individualistic or community bio-genetic uniqueness. We are now splitting the difference and observing the new genetic/ethnic profiles being created in most parts of the world community. Our goal is to attain, in ethnic parallel, a universal high intellectual quotient for abstract thought and education.

Thus far the physiognomies in the World Society are tantalizingly diverse. The new ethnicities forming in the newly modernized societies all over the world are equal in educational accomplishment and now heavily influenced by international communication, knowledge and experience. They are also developing a pride of place and respect for their history. They may not know where they have originated genetically, but they all are now, several generations deep in time, dealing with the history, geography and ecology of the community into which they have been born.

We must treasure both the old ethnic information from the past plus the new ever-evolving ethnicity of our future. Never in the history of civilized literate people has there been an ethnicity that has, so to speak, been 'stuck in the mud.' All ethnicities change in language, style, the arts, technology. We need diversity in our world to protect us from the tyranny of the whole. But no nation, community or international entity can exist divided from itself. That is why we must protect the values of national and community ethnicities as they contribute to the richness of life. The democratic refraction of these diverse values will pour into our international deliberations and thus lead all humans onto a pathway of peace and progress.

There are and will be no pre-established barriers between national ethnicities. Tourism is encouraged worldwide. Immigration is regulated by the receiving entity. No walls are allowed anywhere to bar rational emigration.

D: Slowing Social Change

Galileo could have predicted what would happen to society in the future through his early seventeenth-century experiments with the acceleration of falling objects. The steamroller of social change all throughout the world from c. 1500 finally hit *terra firma*, c. 2050–2070. Mostly the surge of social change was stimulated by the enormous amounts of fossil energy found and transformed into technological/medical uses as it controlled disease and death.

Technology, industrialization and scientific medicine are not merely physical entities. They carry with them symbolic values in terms of the human behaviors that they transformed. And because of the enormous evolutionary differences in human intelligence, this factor is subject to this transformation of our physical lives. The mere physical dimensions of this dynamic precipitated series after series of horrors. Humans could not digest what was happening to them in the transformation of their material existence.

And because of these enormous changes in population, abstract intelligence and the external conditions of life, any stability in the symbolics of social or moral behavior dissolved. There was no conceptual center in the blazing and constant transformation of cultural life. Without rules, patterns, traditions, expectations, humanity slipped its evolutionary leash. We cannot let that happen again. If there is to be a democratic evaluation of choices to be made in terms of how we live day to day, week by week, then we must make conscious decisions in terms of what we want to retain as contrasted with what we need to change.

If the basic values of life—as an American, a Congolese, a Japanese—are dissolved in melted butter, there then is no unified source for rational democratic decisions. Thus, we have to slow social change in all of its diversities so that we can tap what lies deep within our individual and community sensibilities, our acceptance of what it means at any one time to be a rational, ethical, committed (not alienated) citizen of a nation, a people, a world.

Our policy is to allow innovation in technology, economics, social organization to take place on the micro level within discrete communities. If the new seems to work for one or more communities or nations, then it is transferred from Secretariat supervision to a general conclave of our Congress of Nations. There, a two-thirds majority decision to do or not to do is given to the member communities of the World Society.

E: War and Democracy

Clearly, they do not mix. We know this. But it is important to reiterate this fact because there are deep emotional limbic system/mammalian urges to go to war against the *Other*, whoever he or she might be. Very often those who seek power for themselves or their cliques promote wars not for the ostensible right to survive but rather to protect their privileges of power. Rarely do we throw them out in the middle of a conflict.

Certainly the symbols to trip the wire for bloody conflict are there—religion, the flag, and pride of nationality. Name the heroic leaders of the past, and there they are on horseback, with sword upraised, a powerful speech celebrating the dead (Pericles, Lincoln). Could things have been resolved without the trenches of dead, the gas chambers, the cities incinerated by fire or atomic bombs?

The World Society disallows from membership any national, ethnic or community group that wants to retain significant armaments or military capabilities. We allow only a lightly armed police force, which has the right to bear modest arms in the group. The World Society has placed armed forces, highly technological and with advanced weaponry, around the globe for emergency purposes. The mammalian instincts still lie within our arteries.

The Secretariat is always quietly on the lookout for national executives who are raising up demagogic national or ethnic emotions that could spill over into a war mentality. And our Congress can be constituted very quickly to deal with the outbursts of this ancient proclivity of humans. My point is that democracy disappears with the first shedding of blood.

F: Economic Playing Field

There is great potential wealth in the corporate ownership of a steel mill as well as the ownership of a fashion design firm. There will be semi-skilled workers in the steel plant, and there will be semi-skilled workers in the design shop sewing the dresses or suits for the high and mighty.

In the end the democratic question devolves into issues educational as well as economic. It has to confront the reality of the seemingly lowly workers as compared to the owners or managers of the various enterprises in a society. The minimum wage in any nation has to meet the demands of living the middle-class life. To truly participate in the democratic culture of their nation even a person at the bottom of the existing income scale would need to have the economic wherewithal for full participation.

Our histories relate that in British pubs in the early twentieth century, where the industrial workers drank their ale after labor, the political discussions and debates were vocal and lively, the issues well understood. On the other hand the power of these numerous workers was relatively inconsequential compared to the political influence and power of the few who ran the major industries of the nation and controlled the sources of information and communication.

Naturally as work in our world society is increasingly mechanized and automated, we will need ever fewer of the semi-skilled workers, and there will be a widening field of equal opportunity to rise with one's talents. We would hope that more and more citizens will be able to take their place in the economy through the free choice of vocations enabled by an education that results in relatively equal intellectual and economic attainments. Differences in the economic scale will be determined by personality and individual decisions. In terms of educational attainments we will not blame differences in results on the teachers.

It comes down to the issue of eliminating static class structure. Through the generations there is constant movement up and down economically and socially. The nation should make sure that monopolistic controls are lawfully proscribed. The economic structure that a democratic nation or a people subscribe to should not allow for the accumulation of vast personal wealth and the power that this uneven wealth allows individuals or groups to accumulate.

The words, middle class, should mean something in the vision of the leadership and the people of any purportedly democratic polity. This is the goal of the World Society. The Secretariat and the Congress have been regularly called into service to oversee the many political subtleties that our various communities have put together, so as to observe the concrete results of political diversity in its impact on the democratic ideal. Economic equality of opportunity and fulfillment should be front and center in the discussions.

G: Government Bureaucracies

This is an ongoing work in progress, the regulation of the size, power, and democratic political role of those who are employed in government service and those who contract with and benefit from the taxes on the public that government dispenses. Of course these two categories, employees and contractors, pay their share of taxes, but they also claim to vote for and against their own benefits. And what about even the few who now receive unemployment, health, and welfare benefits? Indeed they are today only a small percentage of our respective national profiles. Should they be allowed to vote to influence these benefit payments?

One argument has now been sloganized: 'No representation without taxation.' Still, government employees and contractors pay taxes and can still vote their own interests. Many centuries ago an American president, Franklin Roosevelt, dismissed the idea of federal workers being able to unionize to protect their own benefits and conditions of employment.

The Congress of the World Society has been scrupulous in circulating a wide variety of proposals to the nationalities with regard to the moat that separates the taxed and the taxers in the sense of government regulating but not by itself running the show. At the least the contracting corporations will have other nongovernmental income sources and not be totally beholden to those in power for their various emoluments.

There can be no democracy if government is omnipresent in our lives and dominates the social dynamics. Human dependency on government philanthropy and basic life support is the enemy of a free democratic polity. We in Nairobi received decent wages and benefits, completely revealed to the worldwide public. Our example has been taken up universally by the member nations. Even our established military outposts throughout the continents are minimally manned, with voluntary reserves available from the member polities.

The vision of the 'diminished state' of the Scot Adam Smith in the eighteenth century and that of the eighteenth-century Constitutional founders of the United States (this nation tragically put a hole in its own boat in the twenty-first century) constitutes the philosophical model of our political democracy.

I should state that I have never voted since becoming a member of the Secretariat, being then well into the voting age at twenty-six years.

H: Demography

We are intent on reducing our demographic footprint on the environment for a basic economic reason. With the level of material resources of nature available at this millennial point in time, we yet cannot support the current world population in a middle-class manner consistent with the technologies available to us both in theory and practice.

We also feel the weight of responsibility for the ecological and environmental impact of the gross ballooning of the population of humans as it reached its destructive crescendo in mid-twenty-first century. We need to repopulate our world with as many endangered species of life as we can, so we can go on living with and in nature, not destroying life and ourselves.

There is no democracy in a mass society, merely manipulation, and as it sadly has occurred, manipulation of uneducated, intellectually limited masses dependent on government support, in a world mostly totalitarian in nature. It is doubtful that we can go back to the rural, small townships that the United States philosophers modeled for themselves or the totally independent city states of the Sumerians, Hellenes, Renaissance Italians, always warring with each other. We must try to reduce the human imprint throughout the world so that real discussion and debate are linked to real legislative response by the rational voices of the citizenry.

The World Society has tentatively agreed that given current material, environmental, and historical conditions for our species, we should try to further reduce our world population from the 3 billion plus in 2284, to somewhere south of 2 billion, a century or so from now. Our descendants will take on their own estimate of sustainability. Independent nations should now be no larger than upper seven figures. Urban conglomerates should not be larger than mid six figures, c. 500,000 souls, here including metropolitan areas.

Experience has shown that with current communication technologies, such demographic units are capable of instituting democratic processes, disseminating information, receiving in return full hearings of the will of the populace, ensuring that executive actions fulfill the legislative intent of the citizens, having opportunity for judicial review consonant with local constitutional or legal precedent.

One has to repeat that old aphorism, 'small is beautiful.'

· 17 ·

ECONOMIC EQUALITY

Economics of Natural Selection

The enormous expansion of the human cortex changed the dynamics of *Homo*'s interaction with the material world. Without directive instinct as with lower anthropoids and mammals humans had to devise other techniques to survive economically. We see the use of tools in hominids as far back as a million years ago, a dawning realization that humans would have to 'go out of themselves' to devise means of survival.

Without the usual appurtenances of predators—fangs, claws, large body mass or speed—this corticalized two-legged predator began to shape tools in conjunction with his socialized, cooperative expeditions in search of prey. Clearly it worked, for over many hundreds of thousands of years *Homo* grew larger, spread over the world map and revealed an ever-more sophisticated tool kit for cutting, smashing, puncturing.

Somewhere along this evolutionary time scale, perhaps at the *Homo erectus* level, the tools seem to take on an esthetic dimension, not only practical, but symmetrical, almost idealized. There is controversy here, for what seems to have been shaped for beauty might have been an unconscious result of the process of flaking tools from a stone core. This skepticism disappears with the arrival on the scene of Cro-Magnon man.

Here the tools are without doubt the product of humans who saw in them not only their utilitarian use for economic survival but also as esthetic objects, perhaps of value. Many of these tools carved from bone, ivory, slate, etc. are too delicate to be of any use in the hunt. They are found with decorative jewelry and other exemplifications of the fact that these people had found ways of satisfying their basic survival needs—food, clothing, shelter—and were now experimenting with the concept of beauty.

And we also find amongst these objects materials that could not be found locally where the remains were found. This is an indication of trade. But trade goods are not necessarily the economic items of survival. They are the economic materials and objects of value. Thus we enter into another dimension of the meaning of economics, unique to the human evolutionary message.

Even after the shock of the degradation of our mineral and other resources necessary for the modern economy during the 2050–2170 period of scarcity, we now do have under our international control enough for the basics for all the 3 billion-plus humans on our planet. One hundred years from now as we will near our first goal, a demographic imprint of 2 billion or fewer people, we hope that we will have moved well beyond the danger points of starvation, cold, drought and the inability to migrate away from physical or climatic danger.

There is an inner adaptive force within the human psyche that wants to overcome the limits of nature, wants to reduce not only the threat of impotence and weakness against the mysterious powers of the external world but to master the physical and transform these energies into the mental. We see this in the perennial search for labor-saving techniques and machinery as well as the piling up of security, more and more, even when one needs less and less.

Our belief is that humans translate the need to know the causes of particular relationships in the environment, to understand the structure of the universal causes, the laws of the universe. In our search for knowledge, we naturally endeavor to bring the precarious nature of the human enterprise into our safety zone. This sense of security or survival brings with it the desire to reduce the burden of physical bodily labor. This bodily labor constantly reminds us of our weakness our susceptibility as animal beings. Bring on the machines, the robots, less sweat and muscle pain, more cortical milk and honey.

International Values and Equality

Even as our world began to see the light at the end of the tunnel of horrors and the weather became seasonal again, c. 2150, we asked where we had gone wrong. It was not so much a matter of looking back, but as the new generations began to transition into maturity, the issue was the future as much as it was how to avoid the past.

Our economic theorists saw that the supposed issues of economic equality that led peoples and nations into competition and war was in reality about the dominance of their national values, cloaked in the discourse of economics and power. In the history of those two terrible centuries, the twentieth and twenty-first, the push for power by fascist and communist nations as well as the terrorist bands of Islam seemed to be a basic economic power thrust, grabbing territory and resources. Underneath it all, however, was the boast about the conquering values of the aggressors. Recall the flags, the uniforms, the parades and exhibitionism. You do not necessarily conquer for the gross material and economic advantages. It is always a matter of the ethnic, ideological, or national values embedded in the moves for power.

Now we understand the inevitable outcomes of these various thrusts for economic, military and geographical potency. It ends in disaster for all, innocent and guilty alike. In our new international profile we needed to plan and create a global environment with a wholly different balance of values. In this new historical phase of international relations the old capitalist *laissez-faire* is no more.

The constant jostling of nations for land, guns, people and economic power could not continue without destroying the human race. If a nation or a people wants to puff themselves up in pride, they will have to attend one of Leonardo's famous fabrications in the Milanese festival parades. If not Milan, then we can settle for the more plebian Thanksgiving Day parades of Macy's department store in New York City. No more the flashing bayonets and block-long missiles and artillery pieces on that perennial main street to national annihilation.

There are innumerable economic values that can lead a nation or an ethnicity forward to prosperity rather than war and blood. The tragic loss of populations through conflict, starvation, disease at the end of the twenty-first, beginning of the twenty-second century, closed the door on one epic expanse of human history. The general cultural malaise and political apathy that followed the chaos allowed for the gradual regeneration in the young of a sense that there could be a future for them and their heirs. Why, the thinking went, do we need to live in a world that benefits only a few and for a moment in time? Why does the ignorance and arrogant power thrust of a few have to lead to such an extended period of sadness for so many?

The new generation saw the economic good that had resulted from science and technology. The future being envisioned was not one that mandated new or superseding values for the world community. Rather it was a mandate to tear us away from the social and economic assumptions in all the nations of that intertwined internationalism of the League and the United Nations that had catapulted this human race to the edge of extinction.

Economic equality did not have to bury those drives of humans to master the material environment. It could simultaneously free up the deeper mental values of the arts and culture. What we had to do was to divorce these drives from the assumption of power by one group over another. Indeed, we believed in the diversity, the plurality of values that accompanies the freedom to think and act by any national or ethnic group. We wanted to build into our system the conditions by which these forces of human nature would be so aligned as not to be an economic tool used to aggrandize and debase others, rather, to build on our own values and vision.

Conditions of Economic Equality

In the beginning, for the thirty years or so between 2150 and 2180, a series of wide-ranging meetings throughout the world assayed our economic and social conditions, what had happened to this human species. Here and there suggestions as to how we might go forward were put forth in some specificity. By 2180 it had become official in Geneva. We were going to have to have a real world society organization but with teeth rooted in that capacious and rational skull.

Goal 1: Even with the population decimation during that century of devastation, many knew that the world papulation had to be lowered to what reason told us would be long- term sustainability. This reduction could happen through tragedy or rational choice. We opted for the latter. Mother Earth could no longer supply the quantities of raw materials including fossil energy, metals, radioactive energy to allow for a middle-class population on this planet of up to 8 or 9 billion persons.

Today, 2284 CE, well over a century since these early evaluations, we have some 3-plus billion humans on Earth. Our mediate goal is for a population under 2 billion. We believe that this demographic goal will allow for much greater economic equality of life for all humans. Our technological skills are high today, and our research continues. But there is still much to be reconstructed from that era when our world was in the process of mortal collapse. It will take time to create the technological efficiencies and absorb new ideas into productive industries to benefit the people of the world in the next century or so.

Goal 2: As we have explained above, all nationalities around the world realized that those who survived best after the tide of destruction retreated from our memories, the most intelligent peoples, nations, the most educated, were able to devise a host of survival strategies that allowed them to show the fewest scars. The censorship and the repression of truth had disappeared in the wake of stark reality. We now needed to even out our intellectual potential throughout the world, to eliminate the racial divisions which had harried the soul of humankind. Evidential, factual human equality began to enter our consciousness and perceptions.

Goal 3: Bearing in mind the long-term expectation that the material resources of this planet had to be available to all of its inhabitants, we now needed to establish a new one-world principle. No longer can any national entity, because it happens to discover something valuable on or underneath its soil—minerals, agricultural land, even the beauties of a wonderful ecological entity—be fenced off in terms of its possible use for the members of the world community. The real question involved the sharing of these values with the rest of the world but always giving due recognition of the possible efforts, investment, labor that goes into exploiting these values.

No nation, for example, will ever be able to say to its neighbors, '….sorry these lands, minerals or vistas belong exclusively to us, we'll sell you some of the values involved, but keep the rest for ourselves.' Today, the question becomes how to share what nature has bequeathed to human kind; whether discovered by accident, or exploitable by human labor but never bestowed to benefit only a particular national imprint.

Goal 4: By contrast take the situation of a national or ethnic unit that has the bad luck to be located over a newly activated earthquake incline or is a victim of climate change or any such 'gift' of nature. Take away the possibility of bad governance, national indifference. Wouldn't the world itself have to rise up to a moral sense of responsibility for the truly unfortunate?

After we have achieved a demographically balanced planet we would be obliged to appropriate whatever funds are needed to give this worthy nationality or ethnicity a new start in a new geography or ecology. If this is not possible then the community would be dissolved, relocated, have helped its members find their way through immigration into other nationalities/ethnicities, given the latter's mutual assent.

This kind of assistance would now be possible. The world is emptier. Socio-economic differences are a matter of choice not education or material plenty/want. We envision a world built from wholly new philosophical principles based on our most recent scientific understanding of humanity and the natural world that surrounds us.

Illusion

In the twenty-first century the democratic capitalist societies thought that they could create egalitarian economic societies through growth. However, they did not incorporate the empirical factors which should have been added to the computations for this ideal. Instead they poured out upon our planet a flood of consumer junk which dissipated our natural inheritance. It led to ever fewer percentages of those exploiters who could ultimately flourish through this vision. In the end humanity collapsed into the ensuing chaos.

The egalitarian totalitarians by contrast attempted an enforced economic vision of utopia. This, of course, was a direct pathway to the enslavement of the vast majority and the enrichment in power of small groups of commissars/oligarchs who now ran the camps. Here too the model was wildly off the realities of what kind of humanity inhabited this planet.

The key to both models of failure and the degradation of the species was in not understanding the nature of things but instead, engaging in mythological ideologies that suppressed any hope for the restitution of democracy or economic and social equality. We today believe in the factual strength of our peek into nature's reality and thus into the possibilities of the democratically derived laws which will be placed into action. Only an educated populace that is equal in the essentials of the civilizational enterprise can achieve this goal. Have we gotten it right? Only time will tell.

· 1 8 ·

ECONOMIC PROGRESS REPORT

Geography and Power

Earlier we discussed the inherent demographic constraints involved in the attempt to live with a democratic plan of social life. Along with the limitation on the population of urban and national entities, we also believe that the geographical extent of the political power implicates the economic factor. Nations small in population and size inevitably do not exert the same civilizational voice as the big boys. And thus to keep things in balance ethnically, politically, and economically we have over the past century attempted to shear the political and ethnic power dimension of our international order from the gross geographic extent of each political entity. The economic implications here are clear.

- As noted in the previous chapter the World Society has set concrete restrictions on the economic use of minerals and other natural resources by those who live above or next to the natural resources of the planet. The lessons from the Arab world of the twentieth and twenty-first centuries were clear enough. Their economic power was completely out of sync with their political and civilizational contributions to humankind.

With this unearned wealth they sponsored terror and tyranny around the world. In the process they created an indolent population suppressed by a theological irrationality of the most terrifying sort.
- This unearned wealth in a number of nations as well as the colonial exploitation of weak people by the strong only added to the woes of this world before the fall. The Middle East during the cold and dry decades shrank into the dust, along with their vaunted petroleum and natural gas reserves. There is some fossil fuel still left in these regions. But today the few survivors are interdicted to use them politically and economically to cause trouble. And besides, this radically reduced population has changed in character. They are a small but wonderful population now, ingenious in eking out a bare living from this wasteland.
- Our biggest problem came from Russia and China, each vast in extent but relatively uniform in ethnic heritage. Although at first rejecting new genetic inputs into their ethnicity, they managed to stay close to the educational and intellectual mark of other nations who had to a large extent remake their bio-genetic profiles. It is hard to say whether this at first relatively uniform and immobile ethnicity forced them to voluntarily let go of their ethnic minorities, such as for China, the Tibetan hybrids, and the formerly Islamic Uighurs. The non-Russian, non-Slavic minorities of the ancient Russian ethnicity were also released and now are beginning to thrive. Siberia is now mostly a great international wilderness where nature is reclaiming its heritage.
- By contrast, the United States was a very diverse nation ethnically. In its break-up ethnicity became a relatively unimportant element in national reformation. Today these 'states' have created wholly new ethnic profiles. In this once-great sub-continent of milk and honey large geographies are also going back to nature, the people are prospering once more. Old Canada was always a modestly pluralized political unit of Dominions with significant ethnic/linguistic diversity. This nation easily gave up the goal of bigness. Political secession no longer was a big deal.
- Our debating point for each recalcitrant was that their vast geographies should not constitute an invitation for new ethnic political and economic entrants onto old territories. Rather as was the plan throughout the world, the reduced global population constituted a great opportunity for large portions of the planet to go feral, to return to nature. The natural resources which could be had from these geographies would be

decided by democratic international majorities and monitored carefully for the benefit of all the peoples of the Earth.
- So it has gradually come to be. Great areas of what was once Siberia, Western China, Brazil, the Mid-Western American plains and mountains, the great African tropical interior and the deserts of the north, Canada from well below the tree line in the north, most of Australia, and other areas on all the continents would allow nature to return without the helping hand of *Homo sapiens sapiens*. What has finally propelled this plan forward, albeit with the grudging acceptance of early naysayers, was that the surviving inhabitants themselves had begun to differentiate into separate ethnic unities, especially in China which had a tradition of spoken linguistic diversity.

Economic Profiles

Our hope is that all communities would have a different productive economic profile—some agricultural, some technical, even specialization in medicine, information technology, etc. And there would, of course, be international trade. Any one nation that would discover some new cultural value, as did Nanjiang and Transvaal in the field of medicine during my own time, would have the right to argue for some additional value awards, material or cultural to be added to its economic base. Would free enterprise be limited? It would be allowed as it always has been in human history; free enterprise would exist only to the extent that no cartel or monopoly gouging could be traded for patented value.

- Any national community deserves to be rewarded if it produces something of new international value. On the other hand, we all ask that it share this treasure with fellow humans. Because we are theoretically in the process of creating a human race essentially equal in its competency to live in a modern, culturally evolving world, all human communities have the potentiality to innovate, to create economic and social value for themselves and their neighbors around the world. The trick for our Secretariat colleagues was and is to put this principle into an equation.
- If any particular national or ethnic group cannot or does not hold its own economically to maintain their societal framework, including their tourist or art attractions, then we have a tough situation. This

means that they are continuously on the receiving end of the Congress's banking dispensations of our investment reserves. We would have to ask what they are doing wrong. This could be easily remediable over a generation or two—and through genetic banks if human capital is the problem. The issue should be subject to international scrutiny and expressed official concern. Medical and environmental circumstances and other incalculable factors can also be taken into account.

- If all attempts to right the particular ship of state from dependency on the world outside and to real independence fail, the dissolution of this society must be considered. Its citizens would be requested and then required to abjure from reproducing themselves. Of course, they would be supported by the World Society for the duration of their lives.
- It is important to understand the principle behind such decisions by our Congress. Every national entity, either of a few million folks or in a few special cases as many as 10 million is a mixture of two dimensions of the human persona. One dimension is the material, the discursive element wherein a core of cortical intellectual endowment should define each national entity. Therefore, their economic potential should not vary significantly from any other set of nations.
- The other element involves the particular ethnicity, language, arts, the core industries and occupations that arguably shape our personalities. If a society runs into trouble, a drought parches the land, industry becomes obsolete, or if craft traditions no longer fascinate humanity, we have a Secretariat and Congress to help provide the know-how for renewal.
- Today there are experts and professionals throughout the world who can be called in to evaluate the particular crisis that a nationality or ethnicity faces. It is only after all else fails, if the citizens do not respond given that their intellectual capital remains intact, then humanity has to throw up its hands. We as a world community would need to get to work and neutralize the crisis without generating great pain.

Individual and Community

The above considerations constitute the macro-dimensions of international equality and stability. There is also the very finicky and complex world of micro-economics within national entities. The national always melds into and interacts with the larger world community. The general view within the Secretariat and the Congress is that the rules and regulations that we will enact for the macro and the micro-decisions within each community have to evolve out of the ongoing experiences and innovations of this new world era. We are more obdurate in Nairobi, as is the Congress, about the macro international regulation of the material life of our human community than the pluralistic micro internal dynamics of the nations.

- Prior to my final retirement from the Secretariat another incident put our economists into action. The event also called in our political experts, even some potential military alerts/actions. It all happened in an area of what was old Italy, now a national unit with a population in the upper seven figures and extending over a decent bit of that peninsula's southern geography. At one time it was a quite poor and unruly domain. It now has an evolving genetic ethnography. The citizens, as of old, love their climate and way of life. Not having known other geographies or cultural traditions, even so the altered genetic profiles have done much to reshape the economy of farming, tourism, and crafts.
- Gold was discovered in an outlying agricultural district. The farmers on whose land it was discovered called in family and friends. Soon it became a discrete private mining and refining industry. But of what use was the gold if stored away on the sly? So, they secretly brought in some of their craftworker friends, and the gold was soon in the hands of tourists. The local authorities got wise and quietly skimmed off a share of the profits. Even the national authorities seemingly were in this supposedly secret economic uplift, as they soon lowered the national tax rate. But tourists do travel and ask questions about 'where, when, how much'. Soon Nairobi was tipped off and investigated the situation.
- The law provides the principle that every nation should approximate the rule of equality of condition. As of now we have four categories of national 'value' prosperity within which we adjust our tax and equality schedules: 1. upper middle; 2. middle; 3. upper working class; 4. working class. This particular nationality in terms of GDP and mean income hovered around the edges of working and upper-working class categories. Individual and family income and capital averaged out at a certain median level relative to all the other nationalities. As such their economic place in our international firmament directed the World Society to have them contribute a relatively modest tax remission to Nairobi.

- In the economic structure then existing in the World Society, these taxes were in turn apportioned, locally, nationally, internationally in terms of the specific national condition of economic advance and economic choice. As universally applied, those individuals or families in this national category would be taxed as category 3. If individual, family, business incomes would rise above the low rate maximums, they would be required to pay taxes of over 75% on the additional monies earned.
- However, when an individual, family or consensual corporation strikes it rich, so to speak, they do have the option to take the wealth that they were accumulating beyond the maximum level eligible for the low tax rate and place it, untaxed, into a nationally and internationally supervised foundation, still to be managed by the lucky ones. We are still working this issue out. Clearly the foundation would employ persons and add to the wealth of their nation and international society.
- Further, the foundation would thus incorporate the interests and the passions of the benefactors, thus diversifying the local, national and international economy away from the governmental bureaucracies. In general these philanthropies serve a public good that might not have been recognized by bureaucrats such as myself. Naturally, the international and national authorities supervising these foundations would be on the lookout to make sure they do not become family rackets or be used for other nefarious objectives.
- In the case mentioned above it took the threat of force to persuade the parties involved to get with it. Gold is an integral part of the monetary system that the World Society is developing. It serves a universal good. If you find some, you can keep some of it for yourself and your community and also serve the world at large to maintain peace and prosperity. Naturally, those suddenly wealthy south Italy persons and authorities had to pay up and were in addition fined.

Long-Term Values

To create a universal culture of relative equality, we need to establish in the minds of citizens a philosophy of social life. We want to encourage creative progress. We realize the perennial existence of those inner human drives to achieve goals that go beyond what exists today. And the material and economic domains are ubiquitous foci for this energy.

- History reminds us that civilizations tend to descend into that ultimate corruptibility of plutocracy and oligarchy in their middle and final stages. Over the millennia such hierarchies of wealth have served humankind badly. They result in the kind of revolution in which heads are lost as in France at the end of the eighteenth century or during the firestorm of the burning of the McMansions in the United States as the tide of twenty-first century prosperity receded. The curtain always rises to reveal masses of angry, impoverished humans lurching toward revenge.
- Medals or statues alone will not reward those enormous energies pulsating in the human search for material achievement and ego enhancement. Persons want to show the folks back home and those around the world what 'greatness' is. Now the ethical injunction says:

….take the rewards allowed by your community, be it a working class society or upper middle class. In the latter case your fellows will allow you more private low-taxed income and wealth than in the former. In either case create a rational philanthropic institution for the good of your own nation and the world. Remember to affix your name in gold letters. Live like a human being next to your neighbors but without high walls. Luxuriate on this crest of modesty, and you always will be honored.

Will progress be stifled under such conditions and in all historical settings? We believe not. Perhaps, social change will be a bit slower. We will sleep more easily while considering a wonderful idea or plan. In this way humanity will be able to more clearly see its consequences. Fortunately, at this point in our experiment mega-wealth is scarcely found. So too, we probably have several more centuries to go before we see our goals of demographic balance, plus an economic balance sheet in the profitable black, always as protection against tychistic happenstance.

- We have yet to achieve what our psychologists and philosophers believe to be the most optimistic mix of intellectual levels and personality characteristics in our people. We are essentially dealing with many great mysteries of our human nature. The thrust for economic wealth and power has long been ingrained into our souls. Can we discipline this drive and turn it towards the rational good of humankind? This is the question we continue to ask.

· 1 9 ·
PERENNIAL RELIGION

Whence Religion?

As I write, enclaves of the old religions functioning as integral communities and nationalities still exist. I will comment briefly on a number of remnant religious holdovers as well as the cleansing dimensions we have achieved of what once existed as a heavy weight around the body politic of rationality and reform.

Hinduism: Many of the old ritual traditions still exist in old India, a great sub-continent now divided up into independent nations/communities. These are the old Hindu areas. The impoverished lower-caste masses were devastated during the terrible century. With the rebuilding, the upper-caste survivors and dominants eagerly encouraged a genetic revival amongst the lower-caste survivors and this with mostly Europoid genetics.

Today the religious traditions throughout the new India are ethnically and historically Hindi. The physiognomies of the two ancient indigenous communities are in sharp contrast. The upper-caste dominants rarely conceived of a need for external genes. In terms of intellect and enterprise they essentially equal the international norm, but culturally/ethnically still reveal differences in their styles of endeavor.

Islam: Muslim Pakistan and Bangladesh are no more. What was Pakistan is now a passionately secular area of communities, most of them in the categories 1 & 2, high-level economics. The old ethnic groupings of this once densely populated nation have broken up into a variety of highly skilled vocational centers, ethnicities still shuddering from the memories of religious and ethnic annihilation of their lower classes.

What was once Muslim Bangladesh is mostly underwater now, the monsoon rains have returned, the northern hills bereft of trees are now washed away into a few small delta settlements. The fishing industry is well supplied, and the people, having experienced enormous genetic exchanges, are very different in physical appearance from their former residents. These settlements are under the political supervision of several Indian nationalities, even as the former Bangladeshis are now quite East Asian in visage.

Interestingly, as our anthropologists have observed there is little real poverty, instead a small and a seemingly happily integrated ecological lifestyle. The people are not rich, having a basic category 4 economic status. The fish have returned; the boats are technologically modern. Today there is even a historical and contemporary literature that fascinates around the world. Their representatives in Congress give off a sense of religiosity in their approach to issues. They report no new religious architecture or clericism at home.

In terms of the earlier international religions, it appears that Islam has receded most radically from our cultural scene. Here and there in Southeast Asia, in the islands that were once Indonesia, there are modest buildings that serve as mosques. My understanding is that these island peoples have participated in the new genetic amalgamations to an extent. They are a practical, agricultural people and do not ask for World Society handouts. These are areas with very small populations as compared to that of two centuries ago. Regretfully there was then no external assistance to save the earlier residents from the effects of economic decline, conflict and climate change.

What was the center of Islam in the Middle East is still a veritable desert. The oil ran out long ago. The wars of Islamic sects against one another, against Jews and Christians finally resulted in massive genocides. Then the world suddenly turned, and the Persians rose up to overthrow their clerical overlords. New genetics entered the region with the gradual international renewal. The oil had, by then, nearly disappeared, and the climate of extremely cold desertification saw the holy cities of Islam become withered memories.

Our history books today are full of the tragedy of Egypt. Here a once-great civilization became the symbol of the wasting away of this pathway of life/belief. The disappearance of the supplies of fossil fuel was the first hit to their economy. When the climate brought few rain showers over Africa, the Sudanese and Ethiopians in control of the tributaries of the Nile, in panic, began to build a series of makeshift dams to complete their control of the disappearing rivers. The Nile virtually dried up; terrible fighting ensued.

What was once a nation of over 100 million people, citizens, virtually disappeared by the beginning of the twenty-second century. A few Coptic Christians were allowed to enter into the European nations. The latter's navies sank the rest. No more Muslims were allowed to land on the northern shores of the Mediterranean. The rest of North Africa suffered equally.

Judaism: So too disappeared Judaism and the Jews, they who gave the ancient and modern world so much in intellectual life. Israel, a nation once with a minority of non-Jewish citizens, found itself towards the end of the twenty-first century entertaining a minority of Jews, most having left when they saw the surge of poor Palestinians overwhelming them. Israel as a geographic entity subsequently lost most of these Palestinians to war, terror, and starvation. Since then the Jews, internationally, have more or less assimilated, the educated majority having married into other ethnicities or have been generationally absorbed into the secular woodwork.

The Judaism of the synagogue had become a sterile ritualism overseen by clerics (rabbis) who had mastered the liturgy but little else. Jewish intelligence and creativity made a secular contribution to a world that had accepted them but only for the moment. The *Holocaust*, as enemy of high Jewish intellectual life, first destroyed European Jewry. The waning of international prosperity dispersed most of the others, a people once more alienated from their roots in Israel, North America, wherever. As the memories of the old traditions faded, the few Jews of the new generations dispersed into a variety of ethnic valences.

Christianity: Along with the cultural Hindu ethnicities of the nations of that Asiatic continent now with many languages and practices, Christianity alone seems to have retained an international flavor, albeit with a relatively small footprint. There are churches all about. But there are few observant citizens of this world having a religious connection to the supernatural. One will find very few similarities in the services, the art, the ritual and gospels of these contemporary Christian constituencies. They are a contemporary exemplar of that ancient Protestant pluralism, but often without the traditional New Testament passions.

They survive because of our deeply ingrained humility, our acknowledged sense of human weakness. Also important in this retention of the religious is our universally acknowledged ignorance of ultimacy. Most of these Christian churches are not attached to a particular ethnicity. Each of them constitutes an ethnicity in the making. Many of these communities of belief throughout the continents speak only English. In some cases they have even adopted the liturgy of Islam. It is interesting, if not exciting, to still hear in a few exemplars, the polyphonic choruses of the late medieval period.

Peace and an economic life barely above category 4 of the lower working class is all that they aspire to. No regal pope or international institutionalism has grown up between these sects. No Episcopal economic aristocracy undergirds any competition with the secular world and modern economics in these vestigial homelands. Interestingly a large proportion of these Christian church assemblages are located in the Northeast Asiatic domains, formerly Chinese, Korean, and Japanese.

Most astounding, and in spite of the very questionable leadership origins of the Mormon sects, a number of them are still found throughout the old United States. Now centered near the town of Buffalo, close to their place of origin, they combine a strict moral disciplinary code along with a surging entrepreneurial economic spirit. Virtually all of the Mormon centers are rated in the category 1, upper-middle-class affluence. All of the above groups still identify themselves as Christians—this in the twenty-third century.

Why Religion?

This question needs to be asked because we observe in many of our nations and ethnicities today the seeming resurgence of religious practice, if not its theological underpinnings. One notes many of the characteristics of the old-time religious ideals in national patriotic assemblages and festive holidays. Examples are the solemn speeches, which frequently originate from 'sacred' texts that begin the proceedings and the banners and rituals of proper procedure adhering to historic custom.

There are religio/ethnic decorations, emblematic of seemingly treasured memories from the historic life of the community. Also, certain individuals are chosen for their special moral, ethical or cultural insights, their understandings of the proper ways of living in the community. Interestingly, few political leaders are seen as proper representatives of these events. There is always ethnic art and music, some classical, some folk, vocal and instrumental music, oratorios, that seem to echo those deepest feelings of the heart and mind. They do remind us of the faith needs of humans.

Modern humans, *Hss*, arrived here on Earth bereft of the instinctual armor of hundreds of millions of years of evolutionary success. A huge brain explored a world that they had, unaware, stumbled into. What was it all about? this brain asked, How do we make our living and with what physical aids? No, there were no handy claws or canines to help provide the next meal. The cortex had to figure it out. In reality, there was an external world out there.

Our scientists have not as yet established the theogony of the Cro-Magnons. It could have been as simple as a system of symbolic artistic worshipful renderings of the animal life around them from which they drew their sustenance. We know that the horse was preeminent. Cro-Magnon brain power was conducive to a settled, non-migrating life, the creation of villages and cities. But they must have wondered, thus, the eventual appearance of the gods, edifices of worship, the entire panoply of religious institutionalism.

The gods seem to reflect the aloneness of humans on our planet, the fact that we have nothing to fall back upon in terms of permanent behavioral responses that could guide what we should do in those physical dimensions of life. The gods symbolized this seeming weakness. We humans needed to root our behavior in some higher principle. We needed not only the reassurance of power, that they/we are OK, down here but also the moral sense that we did right.

And it is true, isn't it? What do we really know about the universe, the ultimate powers that are arranged against us or, for that matter, for us? All that we know about ourselves and our fellows, about the world outside of humanity is how we should react to this plethora of sensory information that barrages our brain each second and then passes through our evolutionary derived sensory system. All of our knowledge is a secondary conclusion, a hypothesis constructed by nature's brain.

So, we don't really know anything for sure, objectively. It is all filtered through our biological system. Our knowledge of what is, what might happen, really constitutes a look into the mirror. And of course, as we think historically, our conclusion has to be that we are interpreting things incorrectly. Nature did not put us on this world to self-annihilate. That ignorance is our mandate for conceptual realism. On the other hand the very fact of religion and its gods tells us that humans don't trust the knowledge supposedly conjured up by its secular establishments.

It is much better to sooth the travails of our inner fragility by worshipping powerful images that will calm the fearfulness, our sense of weakness and ultimate insignificance, by appealing to these unknown forces up high, out of sight, beyond responsibility. How often have humans put their ultimate life expectancies on the line in prayer that the higher gods or God will help them survive a critical moment in their lives?

When the worst does occur, do we blame the gods or reject them? No, we excuse the gods or rationalize the tragedy that the gods have not helped us avoid. We go on, again pleading ignorance about the order of things. In our suspicion of secular knowledge we continue to wonder *why* the gods did not respond.

The organized religions of the past did surround themselves with ethnic symbols, a language of prayer, sacred written documents, songs, awesome religious monuments often of great technological significance or of esthetic value. It is almost as if the clerical caretakers of the religion are telling us:

> Look folks, we can't assure you of our secular powers to win a crucial military engagement or cure you of your illness. However, we can symbolize supposedly great miracles of your faith and surround you with powerful musical and visual symbols of faith that will fill your soul even in those darkest moments of defeat and hopelessness.

So we gather ourselves together in social unity surrounded by an external symbolic structure of meanings and allegiances. It is all a means of mutual protection, mutual conversations of perplexing ignorance and weakness. But we have to have an overseer, someone or some group of our fellows who are wiser, purer, who might possibly represent the powers of 'being' outside of, above and beyond our sensibilities so as to tell us about the ultimate way of things in our universe.

We can then gather around in piety, expressing our mutual deference to the sanctity of the place wherein we meet the gods. We become humble and try to do good, ultimately absolving ourselves of the responsibility for our failures. Perhaps our religion helps us to think we can ward off punishment by the higher powers for our actions, win or lose.

Seductions of the Mind

Religious traditions once formed with panoply of sacraments, colorful robes, written documents of the founders, holy prophets, require belief. Religions thrive when there are few secular empirical rejoinders to the structures of acceptance. If, as in the Enlightenment, new knowledge literally blows away the literal claims of theology and power, the door springs open and the followers of the old religiosity flee, often into the arms of a new religiosity.

We note that in our own time the worship of ancient gods as monitored by their human guardians and teachers still exist because of the complexity of the new. Science has opened up our world to a knowledge which has peeled away, for the educated, the claims of a higher intervention into the ordinary affairs of humans. Sometimes people worship that which they do not really believe in. The hold of a tradition on their social and cultural allegiances requires church, mosque, synagogue involvement and support.

Here we enter a new domain of extra metaphysical or ontological belief. We 'worship' in the halls of these institutions, give lip service to the sacred roles of priests, rabbis, mullahs and ministers. For these masses of humans, behavior is not dictated necessarily by religious (theological) pretensions, but rather by tradition, community, often political opportunity. It is only when the individuals involved challenge scientific knowledge as they once did with regard to our evolutionary and medical sciences that they again enter that mysterious domain, the abductor of our rationality, theology.

Secular Ideology

There is an even more nefarious dimension of the religious sensibility for the progress of reason and science. This is the tendency of secular humans to invest in quasi-religious political or philosophical ideologies. Ideologies are sets of belief in statements or documentations of truths about the world around them. They usually are garmented in the cloth of a political party or movement, supposedly formed to rescue humanity from the evils of the status quo powers. And they are almost always at war with the factuality of our experience.

In the late nineteenth century and then most tragically into the twentieth and twenty-first centuries, ideologies having religious import for the masses who had abandoned traditional religious commitments were labeled fascism, socialism, communism, alluring lures for the religious mentality. Then, when these drowned the masses in an ocean of blood, other ideologies seemingly more benign but just as sinister in their impact on rational social and political planning took their place.

These latter twentieth and twenty-first century ideologies grew out of the enormous social changes brought on by science—better technology, industrialization, modern agriculture and medicine. Also, add the enormous flow of humans into the cities, the ever-growing disparities in wealth and power accumulated by small groups of individuals at the point of innovation and opportunity. It all conspired to again magnetize masses of humans into a move, even a populist force, against supposed evil.

The two faces of secular ideologies—totalitarian and social democratic—conspired to generate two terrible world wars, innumerable genocides of millions of innocents, always demonized, accused of being at the power crest of inequality. It was a darkness of intellect that eventually in the late twenty-first century caused the entire edifice to collapse upon humankind.

The Future of the Religious

So, the question remains as to how we can integrate all those very understandable religious motivations of humans without giving up rationality, scientific and experimental thinking about the real physical world—matters of land, sea, sky, and humankind—and without then falling into the insanities that were the secular ideologies of our millennium.

One solution the World Society has exerted itself to achieve is the intellectual unification of our species. We now are approaching a point in evolutionary history such that we are not divided into two human sub-species, one struggling with the other for the ownership of their adaptive places on our planet. One subspecies was outclassed by the other's symbolic advances in science, the various intellectual and artistic realms of abstract expression, an economy of specialization and theory.

One of these competitor sub-species ultimately enlisted the help of ideology and the territorial rejection of reason by some fairly intelligent power seekers. Rump intellectuals warred for generations against their own. If the elimination of the best could not be achieved without destroying all of humankind, then they would suck out the wealth that the productive minority had created and throw it helter-skelter into the provinces of the hopeless. And of course, they grabbed a large fistful of this wealth for themselves.

The various worldwide genocides—against the Armenians, Kulaks, Jews, Bosnians, Eyeglass wearers, landlords, Igbos, Tutsis—constituted a war against the progress of the species and its evolutionary drive towards the emancipation of the body and the fructification of the creative mind. The avalanche of humans and the dissipation of our energy supplies which undergirded this demographic catastrophe in the first place eventually destroyed these multiple insane missions. Fortunately now, with a weakened group of humans on the planet and the birth and maturity of a few wise humans of the new generations, the cause of this detour in human evolution has been diagnosed correctly. And by chance we may be on our way.

The religious dimension of human thought and feeling is perennial. It tells us of our ultimate weakness in predicting and planning for our future. Institutionalize and symbolize these feelings of human fragility and aloneness in your festivals, rituals, ethnic communitarian associations. But above all keep these feelings out of and away from the secular search for the understanding of humanity's place in the universe and the programmatics of leading our species towards what will be its forever unknown destiny.

· 2 0 ·
THE FINE ARTS

African Museum

In the year 2280, during the period when I had returned to the World Society in Nairobi for my final service, I had vacation time to use. This coincided with the opening of a new Museum of the Visual Arts in Akan, the new capital city of what had once been Ghana, near the now desolate coastal city of Accra. This new museum was to resurrect in memory the looted and destroyed museum of African arts that had once stood in this general area.

What I experienced in Akan during a weeklong sojourn has shaped my views on the arts, and in subsequent discussions with our own specialists, it is the basis for what I will write below. The museum was built to mark the one-hundredth anniversary of the founding of the city, roughly at the beginning of the World Society conclaves in Geneva. There was enough technological and economic dynamic at that time so that the remaining tribal elders could conceive of and act for their remnant peoples on proposals for the future.

The invited architects, engineers, medical people from all the continents to this museum's construction and dedication were paid in local products, mostly crafts. The foreigners, like many other previous visitors to this fertile area at the cusp of the Atlantic and the jungles of Africa, contributed their genetics as well as their skills. But this was the plan of the wise leadership. Africa, this indigenous leadership determined, would be at the crest of modernity. Educated intelligence was to be the human path forward. And, of course, as the theme of a unified worldwide intellectual polity took hold, invited gene pools, especially of 'geniuses' have catapulted all of Africa forward in competition with the more temperate zones to the north.

In 2280, the museum was reconstituted with what could be returned of the materials looted when the original museum was destroyed, plus contributions from the rich collections of several European and American museums of African sculpted art, heavy in the use of gold, which was once plentiful in this region of Africa. As an invited guest I had access to a wide variety of officials of this city, now c. 100,000 in population, including surrounding suburbs but not anticipated to grow any larger. The input of genetically accomplished humans was ongoing.

Outside of the city were literally thousands of acres of agricultural and horticultural plantations run on scientific and technological standards. They were the basis of the economic advances of the people, as it involved the total production, processing and export of this high-in-demand industry. There was a small college with simple buildings for the basic liberal arts education of the populace plus an advanced institute of agricultural studies with both indigenous Akan scholars who had been mostly educated abroad plus visiting researchers from various continents.

The physiognomy of the inhabitants was telling. One could see the ancient African substrate but, as with myself, also a heavy overlay of other ethnic markers. The leadership always commented on this plural heritage as ancient and ongoing for Africa and the world outside. They gloried in their heritage and the sad fact that the self-enslavement by their African forbearers had led to the slave trade out of Africa in return for some of the skills that went into the beautiful work exhibited in the galleries of this unique edifice.

A building was being constructed nearby which was to house the performing arts. The rich artistic tradition of choral and instrumental music and dance would now add traditions from across the seas. The comments of the leaders of this new nation, their art people as well as the elected political citizens, brought my emotions to the surface.

They said that they were approaching the point of economic surfeit. Their products were reaching around the world. The gold may be gone; the oil in the gulf was still being searched for, now for the benefit of all of humanity. Beyond these once-enriching conduits with their new ecologically and scientifically sustainable economic patterns, the main goal of the community had to be education in the liberal and the fine arts.

The citizens of Akan had to become more than economic research entities. Now that the jungles had returned to us as a great wilderness park for all to treasure and preserve; now that Africa was becoming an equal, socially, politically and economically with all the nations, the Akans as well as the Transvaal, the Igbo nations and the rest, had to cultivate the deeper human dimensions of our biology. This lay in the arts, the cognitive arts that could flow from our blood and be cortically expressed through all the senses of *Homo sapiens sapiens*.

Prosperity and Art

It did not take much time before the Cro-Magnons stabilized their economy, learned the best patterns for the hunt, the fishing, the gathering as they rolled into Europe, probably up the Danube, north and west, c. 45,000 BP, to 'conquer' Europe and send the Caucasian Neanderthals into extinction.

Almost immediately we see the results of life as ordinary. They had sufficient food, animal clothing, shelter in caves, and fire to cook with and keep them warm. They also were passionate about art and beautification. And we know the results for their approximately 30,000-year flourishing, much of the time barely south of the leaching ice fields. When the hunt no longer exhilarated, they took with them the carved and pottery Venuses, sometimes of a size to be hand held, always a reminder of the welcome beckoning in the return. These were not mere porno evocations of good times. They were beautifully fabricated, most often abstract artistic envisionments.

Achieving great things in the material world, whether a mastodon kill or the wherewithal for a Maserati deluxe, tend to become ordinary once schieved—the hunt, achievements at work, even the object of desire, food, 100mph on the Autobahn. At all levels of cultural advancement the mind searches beyond sex and material objects, for beauty. It is impossible to explain, at this point in our knowledge what it is in the neurology of the human brain that searches out beauty with all our senses—touch, smell, taste, sight, sound.

We humans create beautiful, charming, exciting things from the world that our senses inform us about. And this creation of enchanting, wonderful things is endowed with form, abstract envisionments, never mere reproductions of what our senses offer us. Our minds immediately shape the sensory impact into cultural symbols that have meaning to our community, sometimes to our specific community alone. And that is why we never have to worry that art given enough time and leisure will ever be reproduced in the same way by differing societies, cultures, nationalities.

We can appreciate the different styles of visual art that flowed between the tribal peoples in this Upper Paleolithic environment, from 45,000 to about 12,000 years ago when the ice retreated for a while. Many of these Cro-Magnons moved south from the suddenly colder winters to the Mediterranean, the rivers of Egypt and Mesopotamia. Some scurried farther east to the Indian sub-continent and the river valleys of China to build different 'tasting' civilizations. But always, whether in the Sumerian towns where even the ziggurats had inlaid stone designs, or in Egypt, India or China, as soon as the basics of physical survival were modestly in hand, the art flowed out from the mind into physical reality.

Who can forget that even in the direst periods of the Peloponnesian War, when Athens fought against its brother, Sparta, Athens did not stop work on the sculpture and architecture on the Acropolis—hard labor on recalcitrant stone, mightily pulled up to this rocky prominence. In the worst of times, humans don't need mindless entertainment although that is what some cultures have produced in their waning moments, The great art on the Acropolis was meant to remind the citizenry in perpetuity of the significance of what they are struggling for.

The search for beauty is thus not merely a sensory or sensuous satisfaction, but in its various formations—here we also include the word, poetry, the novel, the drama—it tells us of the intellectual and abstract values we are attempting to protect and nurture. These words are only a halting effort to explain the hold that the arts have on our social minds but they express the centrality that our new international world places on all our national entities for the production of the art that is within the mind of humankind.

To Understand the Esthetic

The question is asked of the Congress of the World Society: Can almost unlimited freedom in the arts take the place of greatly restrictive freedoms in the areas of economic and materialistic development? Much depends upon our being correct in this assessment as to how our international polity will work for us, for the betterment of other species and the planet within which we live.

We have to consider the special evolutionary conditions of the human advance into sapiency. As the brain grew in size, churning nervous energies poured forth. Evolution witnessed a cortex that expanded orthoselectively beyond the boundaries of traditional natural selection even while working on the ever-important and useful practical exigencies of life survival. We humans also have a limbic system, an allo-cortex, the ancient mammal/primate assemblage, genetically linked to the growth of the thinking brain. These structures expanded in tandem, in energetic potency to a point where they have sometimes overwhelmed rationality. The result is that while we are a highly intelligent animal that can anticipate and relate (cause and effect), we are also a frenetically charged emotional primate. The flood of perceptions that enters our nervous system really challenges our powers of mental/psychological self-discipline.

Much of this sensory input is first channeled into the more practical dimensions of earning a living. So too, were these percepts absorbed and tied emotionally and cognitively to the religious, social, sexual rituals and the creation of myths, magic, the sacral, the feared and tabooed. There are thus many ways in which the emotions of our mammalian heritage and the cognitive hominid ordering of these symbolic cores of meaning are finally expressed in overt social institutions. The interaction between what our senses absorb from the outside world, those energies that well up from below, and the final cortical interpretation and shaping of these perceptions, is ultimately what creates the esthetic product.

Critical to the process of producing a beautiful textile, a lovely tool, a song of haunting feeling, is cortical organization. Certainly, a people threatened often will have neither the time nor the energy to entertain new art forms. On the other hand, ossified traditions, fixed in place by religious or political forces will over time dampen creativity. The Jews, in prosperous sixth century BCE 'captivity' in Babylon, sang their laments for the memory of Jerusalem to the accompaniment of lyre or harp. And the host Chaldeans came to listen and ask for more.

This logic of perception is different from the logic of experience expressed in writing. Poetry translates the spoken language into expressive rhythms, music of words and evocations but not into instructions or directions. Esthetic forms build on an inner logic of appreciation inherent in the perceptual materials themselves. Thus the spoken language itself hints at this verbal emotionality, now written down in suggestive form.

The scent of flowers and interesting food can transmit different sensory meanings. For there to be an esthetic experience we need to objectify such perceptions into perfumes, a gourmet feast. We today have a variety of communication forms, even old-time printed newspapers and magazines that try to estheticize ordinary daily experience, wine tastings, flower arrangements having both visual and olfactory impact. And what about the implied beauty of ever new styles of women's clothing? Usually words can only attempt to describe the indescribable.

Relativity in the Senses

How do we understand the 'ugly': smells, visuals, sounds, tastes? Are they culturally relative? Olfaction? Out of curiosity, we might ask what we can do esthetically with the smell of gasoline, burning garbage, sewage, chemical effusions. Think of the sensory horrors of twenty-first century urban slums, the cacophony, stench, visual shocks of the city. And yet billions of humans put up with this sensory bombardment. The human mind can adjust. But the civilized mind will flee in repugnance.

Touch is also difficult to objectify discursively. How do we objectively symbolize the experience of soft infant skin or that of a loved one, take marble, cool and smooth, and contrast the feel of a rose petal? Compare the abstractive analysis of touch to the jottings of notes on paper, composing music preludes, to the performance. Consider also the preparatory drawings of a painter; the architectural and engineering plans for a great concert hall.

This difficulty is true of the esthetics of taste. A cookbook is a poor hint of the great things to come on the table. Perhaps that is why these three sensory modalities, touch, taste, smell, create the most private, effete kinds of esthetic involvement. And thus they are more closely bound to differences in culture and community, and also relative in being appreciated. Our experiences in the real world will tell us or not, that in different cultures some of our sensory, esthetic appreciations will radically differ in terms of their publics.

What we are thus arguing for is the need to appreciate the reality of hard discursive unities, politics, economics, science, technology contrasting culturally with the softer non-discursive shadings of differences in the arts.

Unlike lower mammals, the sense of smell in humans has been subordinated as an evolutionary adaptive, discriminatory sense as compared to the visual and auditory. In the main sound and sight have become the distance receptors for higher animal adaptation. They have evolved over evolutionary time as the major "time to think, attack, or run" information-giving senses. These sense receptors were increasingly coordinated with the cortex in its integration of information, giving the bearers of these sensing/thinking structures enough time for momentary behavioral adjustments.

Here adaptation for survival surmounted the trans-generational processing of seemingly random genetic mutations, aiming for adaptation and fitness. The brain was a paradigmatic "within-the-generations" means for survival. These distance senses integrated information for brain and behavior, allowing for successful procreation, another day in the sun. In *Homo sapiens sapiens*, these two perceptual modalities are highly amenable to intellectual and thus universal human judgments.

Touch, taste and especially smell, as noted, are culturally sensitive. Smells deemed normal or even enhancing can be repugnant to persons of other cultures. One can love the architectural beauty of the ruins of ancient Cambodia, Angkor Wat, but be revolted by the ordinary village aromas of the local towns nearby. The residents themselves would normally think nothing of these scents. Chinese aristocrats once derided the intense body odors of visiting Caucasoid Europeans. So, too, the first Europeans explorers visiting Africa below the Sahara felt free to comment negatively about the body odor of Africans. And of course, the Africans had their own opinions of the body odor of the Europeans.

The highly personal, non-objective, non-discursive world of smell in humans contrasts sharply with our ancient mammalian heritage in which olfaction was a principal source of information about the outside world. The visual and auditory were secondary. Far back in evolutionary history, the ability to separate out the different significances of smell could automatically trigger life or death behaviors. Even today our Geneva and Nairobi poodles articulate their world, without words, by sniffing and burrowing their way around those avenues.

· 2 1 ·

OUR ESTHETIC VISION

Art Fits Our Future

The fundamental question that any national or international plan for humankind must answer is the issue of human progress. How can we maintain the pulsations of the symbolic energies of the human mind while at the same time keeping the world from falling into violence and turmoil? This has to involve social planning on the international level with the hope for continued rational democratic governance.

The history of the human rush to growth and expansion from c. 1400–1500 onward to the twenty-first century is the modern context for our study. Was there even a hint from the mid-twentieth century that we were in trouble? Take the decline of the arts as an indication.

The West was bursting forth with a whole variety of institutional energies during this approximately 500-year period. The growth in power and population was paralleled by a vast unleashing of human creativity. Here, art pulsed with innovation alongside the material and economic.

Never in the history of this human species have we seen so many social, physical, technological and medical changes in our way of life. True, along with this seeming progress blood poured forth from the human corpus. In the end we had never seen the descendants of Cro-Magnon engaged in such terrible wars, many of annihilation and genocide. And these were moderns, *Hss*.

Up until the end the arts poured out, painting, sculpture, literature, the arts of fabric and style, cuisine, instrumental music and song, a vast panoply of the new, often and always of great intellectual depth. We need not describe this panorama of intellectual forms of the different arts and their transformations until something happened after World War II, c. 1950.

Classical music and painting dried up; architecture was suffused with new materials but little vision. Many monstrous engineering and construction monoliths spread over the planet. Most are still there, vast and ugly hulks of iron and cement, once with glass. We still don't have the resources to take them down, now, hundreds of years after their pointlessness. Of course, there was once a point, bigness, power, ego, the wastage that comes with momentary unearned excess of wealth.

Vast numbers of humans were poured onto the world in this era. Its tangibility came upon us without understanding. Now a world connected, the intellectual levels had to be causal in this decline. The arts may have been the first to reveal this sad deterioration in the mental disorder that entered our sensory orbits. Aristotle in his own day decried the degeneration of leisure into 'entertainment.' He had experienced the war-torn turmoil of the fourth century BCE. But he could not have foreseen the twenty-first century version.

The mass media, now electronic, made it easy for those endowed with power and irresponsibility. As a few cynical critics put it, the arts had descended below the neck at best to the diencephalon. The clever, even smart ones took advantage. They gave the masses the entertainment that they needed. After all, with fading material and social expectations, they had to have some form of momentary gratification.

They still talk about similar strategies utilized by classical Central American military dictators. When things got bad and the peons began to wake up to revolutionary talk, the generalissimos opened the cinemas—a free for all, nonstop pornography, and quickly came the street calm. In the connected West, Hollywood and the mass media took up the theme. The beautiful 'gods/goddesses' walked away with wealth, and the masses never knew what had hit them.

This historic scenario has provoked much thought since the first meetings in Geneva, c. 2180. These events of the past stimulated the move to heighten and even out the intellectual character of our species. The arts were first in our minds in terms of the culture that we wanted to find again, the culture that shaped the world from India and China, from Sumer and Greece to the Renaissance and the early twentieth century. These were the eras of humanity's greatness, intellect joined to sensation.

Athens, Florence and Beijing and Nanjiang were busy cities economically and technologically in the classical period of great esthetic efflorescence and population balance. Perhaps we can accomplish this goal given our ongoing demographic shrinkage. With fewer humans spread out amongst the continents, now inhibiting the 'great power' dynamic that has caused so much damage over the ages, less will turn out to be more for our descendants.

Many have argued that the arts are commerce. In those twentieth/twenty-first century days when the higher arts were gasping for oxygen, while the wealth was being sucked out of circulation, governments made their final attempts to save the subsidized lives of the poor and uneducated. Attempts were made to persuade political authorities that these intellectual arts, particularly classical music, opera and ballet required extra financial oxygen to survive. Statistics argued that the arts catered to the well educated and the well-to-do. A bit of extra financial oxygen could mean added prosperity for restaurants, hotels, and transport; tax receipts would also flow. In truth this is not a rational argument for the functioning of a living artistic environment.

Our perspective today is different. First, the higher intellectual arts as they flow freely from creator to an audience of the educated will deepen human social intercourse within each national entity. Second, the plurality of artistic visions within discrete independent societies will encourage appreciative foreigners to exchange and compare values. The mutual fluidity of peoples across borders and the sense of pacific live-and-let-live love of each other's artistic visions will inhibit the passions of war. The impact on the external environment by the arts in contrast to the materialistic gigantism of earlier times will certainly be to the ecological benefit of our earth.

Musical Form and the Cognitive

The development of painting from the early fourteenth century to the mid-twentieth century was one of the greatest cognitive achievements in human history within the visual arts. As such it was tangible, and we can see the results of many creative minds and the dynamics of change from the beginning of this cultural epoch. There is also much ancient and transcontinental art from many time frames to help us view the spectacles of this human urge. Upper

Paleolithic art of the Cro-Magnons from 40,000 years ago tells us of the spontaneous power of this urge for symbolic expression and the art forms that it takes. Of music, we see some Upper Paleolithic wall paintings of shamanistic images revealing practices similar to aboriginal Australian traditions. And we have found Paleolithic flutes which seem to indicate an ancient understanding of the modalities of musical expression.

It's likely that the Cro-Magnon sapients dressed as animals probably were accompanied by simple musical or rhythmic instruments, perhaps even with choral backgrounds. But we don't know. Similarly we have physical evidence of more recent Sumerian musical instruments, pictures of their dancing girls but without the sounds. The Hebrew Bible tells us of the songs of the Israelites probably accompanied by the lyre, but again the sounds have disappeared.

Of all the perceptual receptors, sound that seeks to make music is perhaps the most paradigmatic searchlight into the relationship between our biological energies and their transformation into symbolic meaning. The visual is clearly and quickly the most conducive to abstraction and subject to cognitive control. The three other perceptual realms—taste, touch, and smell—can be refined and cultivated as art forms but with the exception of smell, they evoke few powerful throbbings from our emotional mammalian system.

Music has this connection. Instrumental music, linked to song and dance, digs deep into the basic rhythms of the heart, the soul, and into the libido. But at the same time, music is capable of being organized intellectually, when written in musical notation, whether as a piano sonata in the diatonic form or expressed according to the geometrical teachings of the twentieth-century theorist Joseph Schillinger or Arnold Schoenberg's twelve-tone system. Think too of Johannes Ockeghem in fifteenth-century Flanders composing his complex polyphonic mass, *Missa prolationium* with only suggestive notation for the many vocal lines which the singers would have to interpret and then produce in mathematical rhythmic and notational sequence.

Societal Power of Music

Since the beginnings of civilization and the development of writing, municipal leaders have been tempted to organize and control music, constituting a potential dynamic of expression powerful in its undermining of tyranny. Our historians cite the Woodstock, New York State, concerts in the United

States in the late twentieth century as the beginning of a new cultural and revolutionary ethos among the young. The Soviet totalitarians rode hard to control such creative efforts in all except the exploitation of their folk musical expression.

This was especially true of their classical composers. Intellectuals might have been seduced here but not the passionate young. They were a deeper threat than the abstract expressionist painters. Individual humans remember the songs and that accompanied love and other passions. They elicit constant recall. But the young want the new, not the prescribed notations of the commissars.

Music and language are at the core of every nation's integrity. Civic events have long included children's choruses probably even before the Greeks institutionalized such celebrations. In this civic development of participation, memory and repetition do not suffice. The chorus masters had to devise some kind of mnemonic notation. Whether it be a Tchaikovsky overture to 1812 or patriotic songs of war, music tells us about the life and history of a nation and people. Historians recall that when the Spartans tore down the walls protecting Athens, they did so, ironically, to the accompaniment of hired Athenian flute girls. Did either of these communities subsequently remember the particular tunes that the girls played?

Voice and dance are combined with instrumental accompaniment. In all modern cultures they are already amplified in their expressive power by technology. At the very least voice and dance need only percussive rhythmic articulation. Dance requires costume. Europe celebrated the ballerina. The Sumerians thought dancing to be essential to womanhood. If only we could have witnessed this culture of dance as the Sumerians in parallel created literacy!

The dance always came with music as with the folk idioms and song. And of course in opera and the more popular staged forms, all the arts contribute, sets, costume, personal adornment, and of course music, instrumental, vocal, chorus.

The Hellenes built their theatres to accommodate 15,000 to 20,000 patrons at one time (see the extant theater at Epidaurus), a large portion of any city's population. They included drama, musical performances, and the entire panoply of high cultural aspirations. The Greeks made their arts into public civic enterprises celebrating their history, ethnicity, and philosophical progress. Yet they had their slaves (of their own Caucasoid heritage) to work the silver mines to support both war and leisure.

In one of the earliest descriptions of the Pythian games given at Delphi in the fifth (or sixth) century B.C., a certain Myron is recorded as having received a special citation for his performance on the aulos, a double-reed oboe-like instrument, an instrument used both in Dionysiac festivals and military celebrations and combat. He performed a long series of variations on the mythological theme of Apollo and the Minotaur.

One can guess that this musical tradition was not too far from the improvisations made famous by the Hindu virtuoso Ravi Shankar and his company in the mid-twentieth century. If one travels to the Indian states today, one will hear folk revivals of this tradition. After all, Greek was a part of the Eastern version of the Indo-European languages as was Sanskrit.

This is the whole point of our attempts to create distinct and relatively small political/national entities. They should be separated far enough in their economic and political self-governments (of course, overseen at a distance by the Congress of the World Society) so as to be able to develop a unique ethnic flavor to their social life and their artistic creations.

Populism and Abstraction

During the modern period of the visual arts after 1200, this creative enterprise has been financially underwritten and sponsored for the most part for and by the wealthy and powerful. The medieval cathedrals, great and glorious esthetic architectural creations, invited the community within their sanctified precincts, but the object was control, religious obedience to Rome, to be accompanied by political and economic obeisance.

In the Renaissance and beyond, painting, sculpture and architecture were always linked to the wealthy, whether they were the powerful nobles or the religious, political or mercantile leaders, well into the decline of the art forms in the twentieth and twenty-first centuries. At this point in time they became bizarre objects of ludicrous financial speculation by the same wealthy cohorts.

It is fair to say that the aristocracies were, indeed, intellectually sensitive opportunists who gained power first. The rush of sculptors, architects, painters, from every village in Italy during the fifteenth and sixteenth centuries, the most successful becoming extremely wealthy by the completion of their careers, is evidence of the depth of intelligence and talent in Renaissance Italy and throughout modernizing Europe.

At first the new post-Greco-Roman traditions of classical music were rooted in religious performance. Subsequent to the emergence of art from religious sponsorship, music aided by new technological means—the violin family, the harpsichord, piano, and finally modern wind instruments leading up to the symphony orchestra—plunged into the populace, out from the church and aristocratic palaces and into the theatre and the music hall.

In parallel movements the Globe Theatre and many others in London saw the performances in the late sixteenth to early seventeenth century, not merely of Shakespeare's creations, but also those of Christopher Marlowe, Ben Jonson, and Sir William Sidney. The audiences, some 2,000 every day in London at the various theatres, were largely plebian workers. They understood and loved the language, the poetry, the show. They delivered their hard-earned pence for more of the same.

England supposedly was a monarchy. But populism was then esthetically rooted in the talents of the indigenous working classes. As with music, even the civic artistic floats at festivals in Milan, Italy, many conceived and decorated by Da Vinci, manifested the union of the great with the small, the ordinary folk too, these of high latent abilities. They all showed up to enjoy this intellectual fun.

Remember, as with the English playwrights, Athenian poets, from Aeschylus to Aristophanes, also did not write for the elite. Their plays were the Broadway productions for the average Greek citizen. As noted above with regard to the Pythian Games, the various Olympiads were not merely athletic competitions.

To our own populations, c. 2284, at ever-heightening intellectual and educational levels music beckons. Here the formal principles of the human mind meet in connection with our ancient mammalian and anthropoid heritage. The modern cortex of *Hss* does it all. Our discovery of the formal musical pathways into the emotional depths of humans allows civilization to bind together its universality of expression as well as its national/ethnic magnetism.

Ludwig van Beethoven (1770–1827) was of Flemish and Rhine German stock; he was not a member of the aristocracy. He moved to Vienna to develop his musical compositional and performance talents. Bonn, the city of his birth, was an allied Catholic Hapsburg provinciality. This relatively small community was abuzz with Enlightenment controversy, and plebian Beethoven was likewise affected by this sense of liberation.

His compositions developed from the forms inherited by the Viennese school and became deeper and more intellectual without losing their connection with our mammalian musical foundations. In the end he attracted both aristocrats and middle-class gentry to hear music that rose beyond the given forms of his day to the heights of abstraction and innovation. When he died at age 56, a vast crowd of Viennese citizens attended his funeral. He was probably the greatest creative intellectual that the modern world had seen, literally re-creating what was possible in the world of the arts.

Somewhere in the nations that comprise the World Society of c. 2284 there may be other talents of mind and soul that will fill their own countrymen's hearts and bind the external world even closer with the esthetic glue of a new international feeling of unity and peace.

· 2 2 ·

WORLD WITHOUT WAR

Our Situation

As I write in 2284 in the countryside of peaceful New England, I quietly try to keep in contact with my old associates in the World Society, Nairobi—the young as well as the old-timers. And I do realize that our international structure is a work in progress. The world has not seen such an international experiment in its lifetime, perhaps with the exception of Rome if we accept, as some of our historians claim, that Rome truly was the only real international multiethnic civilization.

I have been part of the bureaucratic establishment in the World Society for over half a century (now emeritus), have studied the archives since its inception in Geneva and must claim that in terms of national or international violence, yes, we must say war, it has been a very quiet time. Thus what I am going to write about in this chapter is what we have come to believe, even given our limited knowledge about *Hss*, is the pathway to maintain our pacifism. Peace on Earth can turn the lifetime of an ordinary intelligent working-class family on any continent into an experience of joyous generational transition rather than the universal pain from which we have only recently emerged.

Predatory Animals

Why violence, why war, why no predictability? We are, like a large percentage of vertebrates, predators. We live off the lives of animals and plants. Precursors and contemporary mammals and other vertebrate predators see the skills of life as embedded in their genetic instinctual signal behaviors. Who they will hunt and kill for sustenance is automatic. It begins as a sensory input. These perceptions trigger a series of autonomic behavioral reactions as the predator's individual behavior sets in and it goes in for the kill.

These signal behaviors cannot be turned off by voluntary decisions. If there are hunger pangs or young to be nourished, the hunt is on and the kill marks an adaptive and selective success. Humans don't have an autonomic system of reflexive instinctual directives to live by. We decide how we are going to live and on what food. We can attribute our transfer of instinctual surety and rigidity, the signal system of adaptive behavior into the symbolic system of long-term decision making, even the individual and group choice to desist from life and die as a suicide to the growth of the cortex.

Only humans can deny this ancient evolutionary imperative, the will to live. This crucial fact alone should make us realize the revolutionary character of *Hss* in the taxonomic hierarchy of the great systems of life. The intelligent forms of life have been increasingly selected out of the vertebrate system of adaptive instinctual behavior for their success in surviving by behavioral changes to external demands, here, within the generations.

This merely indicates that changes in climate, ecologies, geological movements, new challenges to the present order of things, have made us ever more adaptive creatures who can react momentarily to external changes and then be able to bring their young to reproductive potency. The brain and its neurological components for behavioral action thus have marched front and center over the many hundreds of millions of years to become a major survival machine in this sweepstakes of Darwinian selection.

And, of course, one of the main adaptive patterns of intelligent living things was predation, eyes front and center, nose, ears at the peripheral receiving end of information—who, what, where, 'go for it.' Bring home the bacon now and in as many ways as your sensory organs and the brain behind them can muster its library of predatory tricks.

Ultimate Predator

This repertoire worked very well until the coming of the ultimate predator, *Hss*. The mysteries of life are in many ways centered on the as-yet-unlocked mysteries of genetics. We do believe that natural selection is not only based on the phenotypic characteristics, the morphology and behavior of a line of animal or plant life, but also is reflected in the deeper dynamic elements of the living creature. In other words, while natural selection is taking place on the outside, it is also taking place on the inside on the genetics which shape structure, function and behavior.

What happened to the precursors of the hominids, the anthropoids in general, was that the adaptive structure which featured highly sensitized perceptual functions well integrated into a malleable set of behavioral responses all worked well within the reproductive generation of these animals. They were less and less violent predators, having only a few specializations, being rather versatile get-out-of-the-way characters and versatile nibblers of fruit, vegetation, small animals, even a fish if it landed in their laps.

The genes responded too. As they say, success breeds success. The genes were insulated from direct persuasions to what was going on outside. However, the instability of the bio-chemistry in these sets of anthropoid genes and the consequent changeable and successful phenotypes sent messages back to the genoplasm such that successful reproduction of such changes tended to stimulate subsequent and more rapid mutations in the genes, and in the same biochemical direction that success had taken them, *orthoselection*. And of a sudden, some several hundred thousand years ago *Homo* slipped the instinctual/behavioral leash, becoming *Hss*.

It was the genes for brain size that we are basically referring to. Yes, bipedality was a positively adaptive/selective factor in standing straight above the grasses to see what was going on, then to run, climb a tree, or hide in a gully. But the key was a brain that could take in all that sensory information, run it through memory and its internal organization, and then decide how to behave—no more triggers, more likely the thinker, chin buried in the hand in thought. This was an adaptive bonus for staying out of the way.

A Brain Without Instinct

Over the past tens of millions of years the locus of behavior has gradually shifted to the thinking brain from more automatic systems of signals/instinct. But with the coming of the enormous brain revolution in *Hss*, this process has reached a surprising *denouement*. I say 'surprising' because, in addition to the practical eradication of instinctual responses to events occurring outside our bodies or within our minds, we have lost direct contact with that old predatory mammal/anthropoid brain that once held us in such a good position with regard to the selective process of evolution on our earth.

Along with the effacement of direct behavioral and adaptive responses to external stimuli and the powerful supremacy of the grey matter now running the human show, something else has tagged along during this reconstruction. This is the lower brain, the limbic system which sends its waves of energy up into the cortex. *Affect!* As a result we are not merely logical deducers; we are also sexual and sensual seducers. And these waves of emotion can overcome the seemingly logical sources of behavior. Under the flags of nation, religion, ethnicity, even *alma mater*, this lower mammalian brain can send us into paroxysms of emotion and energy.

We humans have become thinking creatures, but we are also under the gun when it comes to behaving in accordance with what our cortex is telling us, of this powerful emotional din that demands expression. The need to survive, to reproduce and bring the young to reproductive maturity now depends on the workings of our cortex to view, hear, plan, and then act. The cortex is also there to harness those waves of emotion which will erupt as we face the tensions of life. We create friends and foes through both our emotional affect as well as our cortical decisions.

Question: how many acts of war are made solely on the basis of a factual cortical decision? How many by wild symbolic emotions, with the cortex paralyzed into inaction?

There is no doubt in our scientists' minds that the desiccation of anthropoid forms, as well as early humans, probably as far back as 8 to 10 million years ago, came as a result of an as-yet-hidden aggressive human presence on the mixed forests and plains of Africa. The passionate search for food, sustenance and protection against highly efficient animal predators was enough to turn our ancestors into very pushy types, 'either your meal or mine.' And the less able apes and then humans who did not measure up to this bundle of thinking and emoting power succumbed to the pressures of *this* presence. There may have been plenty for all, but the emotions of the lower brain said, 'We will have it all.'

In my early days of labor at the Secretariat there was a myriad of discussions and presentations on this issue of human violence and war. In regard to the interpersonal forms of violence—the murders of passion, the rapes, robberies, terrorism, drug and gang killings, much of this taking place in the *favelas* of the poor throughout the world—one view maintained that this kind of violence was leftover instinctual predation by those humans less endowed with a powerful enough cerebral cortex, intelligence, a 'super ego' to discipline such ancient anthropoid and mammalian aggressions.

Statistics reveal that in poverty-stricken locales, the partners of women with children fathered by other men often murdered these children, which is not unlike the way other species of animal males react to foster parenthood.

What about mega wars and genocide perpetrated by the so-called smart ones? After years of argument and decades of programmatic efforts aimed at raising human intelligence on a worldwide scale and simultaneously reducing the demographic injustices of man and nature, these issues have gradually faded from our minds. The conclusion is that the pathology of war and human violence has little to do with primitive aggressive dynamics inherited by our less educated populations. To reduce all forms of violence, interpersonal and national/international wars, we had to initiate a great new philosophical/political program that would be aimed at systemic causes.

I believe I would be correct in saying that the conclusion concerning the interpersonal dimension of violence is that it does have something to do with cortical/intellectual differences among humans from all ethnic groups. What the psychologists call the 'p' (postponement) factor is a product of forebrain inhibitions, the ability to control or internally proscribe behavioral reaction to stimuli, perceptual as well as emotional.

During the twentieth century Sigmund Freud coined the term *superego* to describe a person's ability to develop a conscience. As humans we need a social adjunct to help sublimate and redirect our emotionally and potentially aggressive behavior.

The Big Picture

In terms of interpersonal violence the consensus is that with increased universal education and general middle-class social and economic profiles, a world with natural vistas, air that can be breathed, waters for swimming or fishing, in short the life of civilizing institutions, such violence will largely disappear. And, it has. But remember, it will always exist. To hurt or to kill is an emotional-cerebral decision, and humans are, as we all must admit, undetermined exemplars of life on Earth. There is free will, and violence is a choice.

The big horrors of history have determined our international policies and the structure of our international government. And here, on occasion, is where the friction of ethnic-political dissolution has caused serious debate and controversy. How dare we even discuss the possible dissolution of such human societies? To the historical imagination of many today, small is not necessarily beautiful. This is especially true if you and your fellow humans were once part of a dominating politico-military power of great numbers and geography, and of course, resources—all wrapped up in one fairly homogeneous ethnicity.

But, because these national powers often get into their own trouble and lose much, their power and influence tends to be 'momentary.' The smaller national entities like a Sweden or a Switzerland tend to come through these troughs of pain with some historical humility and perdurance. They have shown us that prosperity and happiness of the citizenry do not depend on having mega-cities or billions of barrels of oil underneath the wheat fields. And so, I believe we are winning that political/philosophical battle, especially with our big patch of little nations now voting a new history into reality.

By committing the many small nations of the present world to an egalitarian view of the physical dimensions of human life, here including the intellectual capacity to keep up with the many scientific and technological modernizations, we insure ourselves against the lure of war to obtain better advantage. We don't know whether the Cro-Magnons as they came into Europe from the mountains and valleys of the Caucasus actually engaged in warfare with the indigenous Neanderthals, their probable cousins from a more primeval time.

There was plenty of room to roam. But just their presence and superior intelligence in coping with the basic challenges of physical survival, their social intelligence in mustering power over the rich lands that they came into, must have frightened the Neanderthals into demographic paralysis, and they disappeared after sharing a not inconsiderable quantity of their female genes with the newcomers.

Today we think that the greatest challenge to peace on Earth and a world without war is a philosophical one. We need to maintain unanimity in our perception of secular reality. We can share our differences in religious nondiscursive ritual traditions. These are cultural issues without international policy implications. However, when it comes to dealing with external reality, we all have to be on the same page, else we will again precipitate the terrible religious and ideological wars of history.

We have studied the tragedy which overcame twentieth- and twenty-first century Islam, then so different than their medieval intellectualism and relative tolerance. The tragedy of the Middle East, in spite of the vast amounts of oil once under its feet which helped to finance the irrational and uncivilized terrorism, was caused by their masses' inability to rise up intellectually to handle the new abstract scientific knowledge contributed by the Judeo-Christian West. Without petroleum to fund their societies they could not even pretend to build a productive scientific/technological Islamic civilization. This is so different from the Northeast Asiatic peoples, who were also late to the scientific party. They had the necessary deep-layered brain power in their ethnicities to be able to match the Europoid West. And when released from terrible ideological chains, they revealed this potential for a terrific dynamic of modernization.

Today—2284

All is different today. Not only do we not have the old ideological and religious obeisance to contend with but racial and ethnic difference in physical features are disappearing, this along with intellectual/educational differences. We are becoming one species, *Homo sapiens sapiens*.

Does that mean that we should not be prepared to meet any warlike responses from our nationalities? No! Early on the Congress voted to put in place small outposts of military units, armed with the most advanced technologies and personnel taken from all points of the compass. These components are mixed into heterogeneous units owing allegiance only to the Nairobi Congress, the democratic symbol of our humanity.

Individual nations are disarmed, with only local police powers. We have had small use of our military, on land, on the seas and in the air. They are few in number. However, their existence has been an effective presence in dissipating the clouds of war and with minimal bloodshed. We are vigilant in ensuring that they should never stray beyond the close supervision of the nations and the Congress of the World Society.

· 2 3 ·
EDUCATION

A Model

In the earliest discussion of the future of humankind, even before the beginning of the Geneva meeting, c. 2150–2170, there was a search for an historical model upon which to build a new order of permanent peace, prosperity and stability for our species. The period after WWII (1945–2000) in Western Europe was frequently used as an example of nations melting swords and engaging in the unification of the euro, trade, education and scientific technology. It seemed to all that the movement into science and open doors of communication created a wholly new international system. This institutionalization of peace and international educational norms eventually brought the communist world down and then included their people in an embrace of this educational perception of reality.

Of course, what followed into the twenty-first century was messy, because of the conceptual errors made with regard to human nature, human intelligence and economics. Because of advances in transportation and communication the entire world was now magnetized into the international system. I have discussed these errors in earlier chapters of my memoir. I would now like to write about our still-evolving views as to how we can keep our educational focus on unity, rationality and scientific progress. At the same time, however, we need to allow for diversity of perceptions, approaches to the rearing of the young, their desires for cultural connectivity with their homeland as well as participation in, traveling and learning from other educational perspectives.

International Curriculum

The concept of an international curriculum is a weak way of saying that we have over the past century created norms of educational rationality which the nations of the World Society must adhere to in order to maintain their membership. What has torn the world apart in the past is precisely the divisive education of the young into secular ideological or religious models which foster not merely hate and separatism but a permanent alteration in any conjoint vision of external nature and human society. This has led to war after war, terrorism, crusades and genocide. Truly the horrors of human behavior begin in the classroom.

What we have striven for is the acceptance by our membership of the secular scientific values of a naturalistic perspective of the physical and the social worlds of our species. To an extent we have developed curricula and texts for the young that bring our membership together into a unified perspective. When we have had problems with regard to educational and philosophical values, the issues were far more serious than momentarily troublesome physical battles between neighbors or perceived adversaries. We have used our military very sparingly in these latter cases of a few months of conflict.

Where it came to a more general push and shove between the World Society majority and several intransigent nationalities, it was the educational practices in the schools and towns within several nations that led to a vote of expulsion. We needed a two-thirds majority of the Congress for this, and we got it. Nationalities in several cases wanted to infuse theological perspectives into some of the social science teachings; in other cases it was nationalism and the hint of an attempt to regain what had been surrendered in terms of land and resources.

In these latter cases they were doing very well as nations before the expulsion as were the other 'fragment' nations that had been separated out of the once-greater mass. They did not have to agitate for more 'lebensraum' or confreres. Now dispatched out into the wild blue yonder with few nationalities bypassing World Society sanctions, they now might have more people, but not more land, and they are poorer economically and culturally. They'll be back.

With regard to the religious nationalities and their theological educational biases that we wanted abandoned, the two nationalities involved came back after almost a decade of expulsion and suffering from the economic and social sanctions that were imposed. We needed only a majority vote of the Congress to bring them back. It is interesting, in that the above nationalistic fragment of a once-greater whole has been a long-term recalcitrant. The dream of ancient glories and power seems to impose a tighter hold on the minds of a citizenry than the theological educational issues that the returnees were once committed to.

Basically our holdout nations have been unable to commit to the fact that our physical life as a species has to be planned out and adjudicated under the larger umbrella of all the peoples of our world. We cannot have them drumming into their young people's minds the supposed glories of the past and the injustice of the present, hinting of their desire to amalgamate with the neighbors they once subordinated. These are now independent members of the greater international whole. No holiday flags of old, songs of conquest, magnifications of the once-bestial into new nationalistic heroes. This old-time garbage pouring into the minds of the young was a strikeout as far as the Congress was concerned. It still is all too ugly.

Our economic and social noose is not tight. But our observation of all nationalistic activities as well as the texts and teachings to the young are continual. And, our military is on the alert. *It is only education, isn't it?*

Cultural Values

Following World War II, West Germany needed to redirect its educational system away from the insanities of the Nazi era to the more distant past when Germans were great contributors to Western civilization. Eventually communist East Germany abandoned its wall of irrationality and was joined with the West.

The Japanese, the Italians, and all others who collaborated in the genocide of humanity have since also redirected their educational systems. Since the reconstitution of our World Society and the separation of nationality and power, we have encouraged all of our new nationalities to mine the histories of their citizens but in the context that they and the world are now taking a new pathway into the future. We are all committed to this multicultural, multiethnic system of values that no longer alienates the young from the outside world and yet gives them the opportunity to enrich their lives with the particulate ethnic values that all humans need and desire.

It is really an educational challenge to balance the elements in our human nature so that we can be at peace with ourselves as discrete peoples and with others having different cultural valences. But overall there must be a unity of perception about the nature of the world we live in, both the universal physical-material elements and the plural social ethnic dimensions.

It is and will be a tough challenge to so balance a world within, having so many new and differing nationalities, all without the opportunity to endow their ethnicity with the energies of conquest and war. But that is what now constitutes the essence of the education of the young, teaching them about the overarching principles of life as well as the discrete values of the home place. Finally, the toughest task of all: teaching and discussing the complex theoretical and factually interpenetrating elements of both worlds.

They still use that aphorism once directed against a failed American president, '...he could not walk and chew gum.' For the sake of our children's future we must learn to value and balance a variety of seemingly contradictory dimensions of our lives. Another old aphorism about the judgment of competing values, in this case literary expression versus blatant pornography, '...we know it when we see it.' In the educational case we will need to perceive and analyze the shadings from true, livable and admissible nationalistic teachings and those which cross the line into dangerous propagandistic practices that endanger the peace.

In the world of reality, there is no such thing as 'everything goes.' If we are Floridians, we can't necessarily be New Yorkers. On the other hand, if Floridians are going to live in peace on the same continent with New Yorkers, they have to agree on the terms. We humans are as yet a mystery to nature, too dangerous to let down our guard. Educating the young becomes the most delicate as well as the most precious responsibility of the old.

Organization

It is quite obvious that in the early age levels when the young are becoming literate and exploring numbers, their orientation has to be toward family, neighborhood, town and countryside as well as nationality. The international order that we are attempting to persuade humanity to enter is yet a highly abstract concept. Thus for the very young, perhaps even through the early secondary school levels, the ethnic side of things must predominate as part of the development of an individual's educational skills.

The question of public versus private has often been raised. The fear is that a system of education independent of government support and intervention can become a haven for the privileged. Here they isolate themselves and take on the personality and attitudes of the powerful and the elite. On the other hand a completely state-run educational system easily can descend into an authoritarian vehicle for the mind control of the people, either for the advantage of a few or to pursue aggressive expansionary policies by intimidating the masses to join in. The philosopher J. S. Mill first taught us this lesson in his essay, 'On Liberty.'

The United States before its political and economic disintegration, and then fragmentation into a series of geographical and now ethnic entities, had a reasonable educational balance of public schooling and a number of fine state universities. At the same time a vigorous private or independent sector of education competed successfully with the public institutions. Neither form of financial support led to one group taking on a divisive philosophical and political/economic orientation. In fact, the religiously founded colleges and universities for the most part participated in the ongoing universalistic higher education orientation.

The destruction of the educational edifice of the United States, along with other socio-economic factors, was due to the gradual restructuring of the public schools into remedial institutions along with the widespread decline of state collegiate institutions. A few held out for international standards in the hard sciences, humanities and social sciences. The latter two areas of study, however, became the Achilles heel of American higher education. Naturally the better-educated social classes who held this society together opted mostly for the independent institutions of higher learning. Only a few of the state institutions were able to insulate themselves from the popularity of the major academic specializations: Hollywood, computer games, rock music, basketball, and football.

In our own day we are as yet too poor to support the luxury of even a modest minority of non-governmental institutions in both the lower schools and the colleges and universities. There have been attempts to rescue the relic campuses of the old-growth era of institutionalized education. The rebuilding and maintenance of these fossils has been a tough row to hoe. All but a few of them are decrepit empty shells in often beautiful park-like settings.

In the long run a mixed organizational structure of nationally run and independent schools seems to be the most stable mix as long as our aims at achieving universal parity of brain power amongst our different constituencies can be firmly established.

The great hope for the universal acceptance of our secular scientific perspective lies with higher education and the trend that existed amongst the best in the early 21st century. A wide variety of modest satellite campuses have in our time been established throughout the world to help the students to experiment with new ethnic and cultural tastes while being educated in a universal intellectual language whether at an Oxford, Hamburg, Harvard, or Tsinghua. This is our goal. I myself profited from the early rehabilitation of the universities that I attended before my life in many different ethnic and geographical venues.

What has happened *vis-à-vis* the major research universities of our world is that the Education Committee of the Secretariat, with the ongoing certification of the Congress of the World Society has made substantial grants to a number of universities on each of the habitable continents with the intention of creating international universities open to all qualified students. These grants will create an incipient class of hard-working internationalists in a variety of walks of life. So far as I am aware, the diversity of grants or their renewability has not been seriously questioned. The competition is great; the responsibilities of our education specialists in Nairobi weigh heavily.

· 2 4 ·

HUMAN SEXUALITY

Perspective

Never before in human history has human sexuality been so divorced from procreation on a worldwide scale. The past one hundred plus years of population reduction have given us a breath of peace and a sense of future reconstruction. We are still far from the goal of the Congress, which I hear is ever-more committed to a global population level, given our resource and economic capabilities, to a figure of fewer than 2 billion humans, closer to 1.5 billion. And this voluntary decision by the human race is completely at odds, seemingly, with the those dark days of war, terror, disease, draught, cold, social chaos, which catapulted humanity down towards the figure where we now are, some 3 billion plus folks in 2284.

Now admittedly, we have in the past century or so, with the aid of an increasingly disciplined population ever more in cognitive control of their aspirations removed about a billion humans from the rolls, through birth control, rarely through sexual abstinence. And this is that great puzzle, outlined in brief in my chapter on a world without war. Humans of high intelligence have developed a cortex that can interdict the limbic throbbings for physical aggression or random sexual relief—here, the result of conscious cognitive decisions.

Is there an animal that can deny the drive to procreate? Is there another animal that can consciously decide to post-natally kill their own healthy child? This should tell all traditional zealots, that there is a new dimension in biological nature which can circumvent several billion years of genetic automaticity. This new genetic intercession, however, is not created by a verbal metaphor which priests, mullahs, ministers, and rabbis once decided to label as 'god.'

New genetic combinations allow us to make such decisions, to kill or not, to rape or not. These genes, however, do not determine us to make a particular choice. Here is that wonderful human capacity called free will. And here also is that mysterious dualism which evolution has produced in human nature, the emotional lower brain which provides the energy juices to allow the cortical brain and its consequent symbolic behaviors to rise to the highest intellectual and artistic levels of creativity such as those of a Raphael, a Bach or a Newton.

Indeed, the human sexual drives exist as they do in related animal species, only *plus*, *plus*. The miracle and the mystery are that they can be, with high intelligence, placed under the behavioral control of the cortex, then released, inhibited, or even as in the choice of celibacy, extinguished behaviorally. And thus with this understanding in mind, national policies throughout the world towards sexuality, as with many other issues have been shaped, to absorb this understanding of *Hss*' unique evolutionary position. Here is vast emotional power that can be poured into human behavior. The cortex, in the end, can and should decide how these powers will be released.

Policy

If we shift our perspective to the issue of children born in each generation, it is clear that society has a great interest in procreation policies. For the most part in a world where the people have democratically voted to reduce the number of children born each year, we have reached a point where the death rate is greater than the live birth rate. Sexuality then becomes mostly a recreational activity of humans subject to a variety of social stability and health regulations.

When we speak of the *who* and the *how* of birthing regulations in nations apportioned throughout the world by a system that attempts to equalize the burdens and pleasures of demographic discipline, we are essentially out of the sexuality business. Levels of intelligence, economic viability to raise children, methods, natural births within a family setting, specialized techniques for approved candidates for various in vitro fertilization and other techniques are all important. All doors are open. We do not see any of these issues—abortion, sperm and egg donations, the entire medical panoply of the process of human reproduction—to be closed to scientific study and possible social adoption.

No spurious sectarian 'right to life' issues are raised today. Our citizens are too intelligent to confuse these slogans with pacifism or religious definitions of sentient life. If issues arise during an approved pregnancy and the fetus is affirmed by our medical authorities as being in a good state, we of course allow the pregnancy to continue until viability. Before that point, as Aristotle defined in his *Politics*, the time of awareness of movement of the unborn, the future mother can decide about a medically supervised abortion. The principle for society and the family has always to be, first, the safety and health of the future mother; secondly, the pregnancy should result in a healthy normal infant, a child who will contribute to his or her own personal fulfillment and be a benefit and not a burden to family, nationality and world society.

There is a bias in each nationality's choice and eligibility to pick the lucky ballot in the lottery, which will say *yes*. Marriage and commitment of both parents to each other and to the raising of children, even if it is only one child, are critical. Our authorities all over the world have created standards of eligibility by examining historical family profiles. Especially as we are in the process of lowering population numbers, the raising of a child now has become almost a sacred privilege and obligation.

At some point in the future when the evidence argues for a stability and equability in the relationship of our species to sustainable middle-class living, conception and the raising of children will become more traditional. Still, the current high standards as to by whom and how children are to be conceived, born, and educated will always be in place. We are and always will be domesticated animals.

Sexual Variations

When we leave the world of procreation and sexual patterns associated with this act for the future of humankind, we enter a very different sexual world. Most fervent is the claim for the social legitimacy of homosexual relationships, even the social sanction of same-sex marriage. Then there are issues of local and perhaps international regulation of prostitution, pornography, transgender issues, and a world of diversity which evolves, retreats, reemerges in our confrontation with the need of societies for rules and regulations.

The World Society Congress has long believed that such nonreproductive enjoyment, relationships, dimensions of sexuality are strictly national in regulation. This area of human life activities might be called the micro-arena of behavior and, like the ethnic culturally pluralistic tonality of international life, ought to be tossed into the regulatory laps of localities and nations.

Should homosexuals have the right to marry? Some nationalities agree and others do not unless children are involved. If adoption is a possibility, one nation would say yes, others no. This option for nationalities to frame their ethnic values around issues of sexual norms seems proper. That would apply to a wide variety of sexual issues, prostitution, pornographic literature in electronic communication, a wide variety of possible sexual patterns, polygamous/polyandrous, and others, today unimaginable to prosaic minds such as mine. These issues would come down to matters of ethnic, cultural, and moral orientation of the particular nationality

So, how to deal with sexual dissidents? Clearly we must negotiate with those who reject the ethnic patterns within the societies in which they were born or where they reside as citizens. The answer is the old ethnic solution. Allow for the emigration of those who can find a home in a more comfortable ethnic environment and who would be received by that nationality. Naturally receptivity to new immigrants depends on the views and conditions of the respective nationalities. What is the state of their population structure in terms of numbers and educational levels? If the individual is attractive to the receiving society then, good luck and *bon voyage* from the old.

Suppose that an individual or individual groups cannot find a recipient nation to take them in? Then, perhaps the dissident, along with fellow dissidents from this one nation or others who want to leave their respective nations for the same moral or behavioral sexual reason, can appeal to the Secretariat of the World Society, placing in this appeal all the economic, social, environmental, demographic arguments for the creation of a new society but also stating the implications in terms of the geographical and demographic impact on all the other nations in the World Society.

The Judicial Committee of the Secretariat of which I was once a member would present these factual, material arguments as well the negative considerations and would present the petition to the Congress for their majority vote yea or nay. There were several yes votes in my many years in the Secretariat.

However, most of the appellants during my tenure were too vague or uncertain in their hopes for long-term national viability and thus were turned down.

Stuck in their originating nationalities they were forced to take their beliefs and life-style underground. If this action did not affect the workings and stability of the nation, the officialdom was encouraged to turn a blind eye to these truly tragic individuals. In several cases of extreme persecution, our judicial committee voted and was confirmed by the Congress to submit a cease-and-desist order to the majority elected executives of the particular nationality.

In other cases the behavioral lifestyle of these deviationists proved injurious to the society and the international framework such that isolation was warranted and confirmed by the Congress. Note that in most cases the issues were so unique that no general rules were applicable, the locales had to make their own difficult decisions as to how they wanted to live together as a nation.

How Far Freedom?

Considering the power of our sexual drives as well as our other emotional lower brain allegiances and alienations, every nationality has the right to strive for social survival. In other words the majority has the obligation to say '…this is our way of life; these are the behaviors we will admit to our culture; and this is what we will prohibit.' Naturally in the international scheme of things our elected *Congress* can say to a nationality's predilections in terms of sexuality and the urgings of the limbic system, 'Sorry, folks, you have gone over the edge of international assent, your decisions endanger our compact of peace, stability and progress.'

Thus although our international focus in Nairobi has been on the larger issues of public life for our species, allowing the nations the responsibility for managing their internal cultural tastes and values, we maintain a respect for a nation's attempts to rein in the seeming liberties of any sexual trend, whether it be prostitution, polygamy, homosexuality, or a seeming anarchic trend in consensual sexual relations which is having an undermining effect on its civilizational profile.

These restraints like the liberties we grant to the private lives of humans are never unlimited. Most cultures have identified sexuality with love. Love is a term of the cortex. It connotes the cherishing of a human being as part of a life which involves long-term commitments. Denuded of its connotations of care and respect and the union of the passions of the brain with the conceptual values of life lived rationally, sex as mere power or casual lust has to remain on the periphery of any nation's cultural ambience.

A social unit's easygoing embrace of random sexual practice endangers itself, its neighbors and the wider world. We have learned the hard way that individual and community freedoms always come attached to a cortical harness.

· 2 5 ·
PERSPECTIVE

Humans Are Different

Intelligence has long been an adaptive pathway for animals' survival. As long as our environment and our ecology are in constant change, instinctually rooted genetic change will not by itself serve as a conservative redoubt against fast-moving intelligent animals. Humans derive from ancient lines of other mammals and anthropoids. Not too long ago in sidereal time, there were a number of such hominine lines emerging from the Ice Age mists. All were significantly in advance of typical monkey and ape anthropoid lines in terms of brain power. But something even more radical happened perhaps as recently as half a million years ago, perhaps earlier.

Nature occasionally glorifies in producing a momentarily successful monstrosity. Deviants, such as the giant dinosaurs that pounded down the earth some 80 or 90 million years ago or those large-antlered Irish elks, some 100 thousand years ago, were all to quickly vanish from this planet. Our own 'peacock,' a super-brainy quirk, was Cro-Magnon, *Hss*, in origin northerly, white of skin, Caucasoid by happenstance. In exchange for this massive cortex and an enormously passionate neurology, nature eliminated any staid behavioral instinctual guidance, the heritage from the past.

In tandem nature genetically dragged along the lower brain with its massive passions. Too often in its surge of energy this part of our brain evoked tremendous urges for sex, war, terrorism, genocide, football, soccer, rock concerts. For 365 days during the year the cortex had to deal with this clouding of logic, forbearance, prediction. But the lower brain also assisted in producing a Shakespeare, a Leonardo and a Mozart.

This cortex was supposed to get us out of any unexpected challenge to our survival. It has largely failed, overridden by our neurological inheritance of emotion and ideology. Both dimensions of our thought, however, have, and, in addition, caused, the spread of both passion and curiosity throughout the world in the genes that were gradually transforming *Homo sapiens* into *Homo sapiens sapiens*. The hope is that the cortex will one day rule the roost and direct us towards salvation as a species without destroying the creative powers that reside in our passionate selves.

Above I have described the condition of humanity in the twenty-first century as it plunged over the cliff in self-annihilation. What I have written in the previous chapters of this memoir is the cortical response of the survivors, a response of the past 100 plus years of international governance. Our objective, was to bring all of humanity not merely to a higher intellectual potential and discipline but to eliminate in peaceful bio-social integration all the various and ancient human lines which connoted physical/racial differences. The most important differences of the human mind, the ethnic and cultural, will be our new redoubt for maintaining the unencumbered dynamic of symbolic change, and passion in this now unifying species of *Hss*.

Our Challenge

So here we are, now cognitively attempting to peacefully reduce the human imprint on our delicate ecology and our waning natural resources. In earlier chapters of these memoirs I have outlined some of our plans for the future, given the assent of the world citizens through their representatives in the Congress of the World Society. The League of Nations failed with 63 members, in and out. The United Nations failed with 193 members. We have far more than twice the latter number. That is why we use a rotating system of voting members in our sequence of Congress meetings.

Our mission is to keep fairly tight international controls over those material and power factors which over the millennia have generated so much pain.

Our control is directed only toward the chaos of change. The new theme will be a slowly thought-out and regulated rhythm of innovation by which we might improve the material well-being of humanity from the technologies of communication to the sciences of health and human welfare.

On the other hand we give a much freer and looser set of guidelines to our member states to experiment and advance in the so-called 'nondiscursive' dimensions of our symbolic life, the cultural and ethnic. We want our most talented men and women to put as much energy into designing a new fashion expression as in creating a long-range airship.

Will these international controls work? Will the allowances of freedom in the cultural and ethnic drain off enough energy and innovative enthusiasms as did the wars and the accumulation of billions in wealth. The prognosis is not good for all overarching attempts, even totalitarian attempts to control and discipline the generators of energy in the human mind. At the moment we are still in the reaction phase of a series of horrors that have inflicted an almost fatal wound on the international human body. So we hope but are not overly optimistic.

The key to optimism lies in the long-term enabling of our thinking cortical brain. Of course, it has to be joined in passionate analysis by our emotional brain. We will then have to factually and passionately confront momentous problems which might suddenly appear.

Finally, the elected leadership will have to put it frankly to the various nationalities of our international constituency. We will not be able to survive if there is not a unified and internationally agreed-upon scientific/philosophical method of dealing with the ever-changing and ongoing flow of new challenges.

The Big Picture

The truth is that there is no ultimate future, no heavenly destiny here on earth into which we will arrive. We must simply keep the old ship afloat and moving gently forward, always against the tides that nature sends forth to throw us off course. We can't go back to the marginal sustainability of the post anthropoid world, hunting, gathering, wandering with the migrating herds. We have wiped out that alternative. There is yet that cortical brain to help guide the emotional energies that are constantly creating ever new worlds of cultural meaning.

As I, Iaian Vernier, in the late fall of life, look back on what we in the World Society have attempted to achieve, rationally, for our children, grandchildren, and now possibly for my own family, a forthcoming great-grandchild, our work has not been insignificant. We are well on our way towards unifying, biologically and intellectually, our species. A structure is now in place to create a basis for economic and social equality. The breakup and shrinkage of the old mega-states will now open many new pathways for ethnic and cultural evolution.

Human nature in spite of so much scientific research into our sociobiological functioning as well as the study of our evolutionary and historical heritage tells us little more than that it is yet a 'buzzing, blooming' mystery. We just will hope for the best.

· 2 6 ·
FINAL THOUGHT

Would the twenty-first century have had a different outcome had Albert Einstein been a sperm donor?
Kyushu Island Delegate, Congress, World Society, Nairobi Enclave, 2255

Also by Seymour W. Itzkoff from Peter Lang Publishing

Judaism's Promise: Meeting the Challenge of Modernity
ISBN 978-1-4331-2529-4 (Hardcover). ISBN 978-1-4939-0871-6 (E-Book).
ISBN 978-4331-2626-0 (Paperback) New York: 2014, 2015
Follows Seymour W. Itzkoff's well-received three-volume series, "Who Are the Jews?" *Judaism's Promise*... confronts the many revolutions that have reshaped Judaism over the centuries, allowing it and its people a path of leadership into the modern world......The book's basic concern is with the withering of contemporary Judaism as a force in contemporary Western Civilization.

Liberty's Dilemma: America, Two Nations, Dependent/Independent
ISBN 978-1-4331-2529-4 (Hardcover). ISBN 978-1-4539-1309-3 (E-Book)
New York: 2014
The growth of a seemingly permanent dependent class on government in the United States has gone largely unexplained. *Liberty's Dilemma*.... points to the declining intellectual capital in large portions of our society as cause. Can "liberty" in our nation so survive?

Humanity's Evolutionary Destiny: A Darwinian Perspective
ISBN 978-1-4331-2545-4 (Hardcover). ISBN 978-1-4539-1256-0 (E-Book)
New York: 2016; Illustrated
In evolutionary history, no two sub-species have been able to cohabit a limited ecology. The outcome of this inherent selective conflict is written in the history of the human genus, *Homo*......Civilization will eventually be shaped by the domination of the sub-species, *Homo sapiens sapiens*.

The Inevitable Domination by Man: An Evolutionary Detective Story
ISBN 0-913993-16-6 Illustrated. Ashfield, Mass/New York: Paideia/Peter Lang. 2000
"Itzkoff takes us on a fascinating tour, beginning with the origin of life through the synthesis of the eukaryote cell.....to the final emergence of *Homo sapiens sapiens*. What he has to say is worthwhile."
Acta Biotheoretica -The Dutch and French Societies of Theoretical Biology

"Though a philosopher {Itzkoff} feels that empirical data must underlie potential answers....his book is packed with extraordinary detail, recalling Darwin in his patient accumulation of data."

ASCAP Bulletin: Across Species Comparisons and Psychopathology Society. World Psychiatric Association

"The author sets a difficult goal for himself to demonstrate that the continuing development of man, 'a highly intelligent, information processing creature' that controls the eukaryote adaptive zone, is inescapable given time and the physico/chemical fitness of the Earthly environment. To do this he draws in such diverse lines of evidence as the vast panoply of fossils and the demographics of genocide....admirable breadth of examples."
The Quarterly Review of Biology'

"Masterful"—*Contemporary Psychology*
"Intriguing and Informative"—*Choice*

www.ingramcontent.com/pod-product-compliance
Ingram Content Group UK Ltd.
Pitfield, Milton Keynes, MK11 3LW, UK
UKHW021846140426
5217IPUK00022B/1614